Perspectives on Energy Poverty in Post-Communist Europe

This book explores the issue of energy poverty in post-communist Europe and shows how it is viewed and addressed through public policies.

Energy poverty is severely affecting many parts of the European Union, but up until now only a few comparative analyses have been developed to understand the phenomenon and its diversity throughout the region. Filling this gap, this volume focuses specifically on the Eastern European region, drawing on contributions that cover a wide range of countries including Germany, Hungary, Poland, and Romania. This region has undergone significant transitions over the past three decades, but, as the contributions demonstrate, it still faces major challenges to providing clean and affordable energy to its citizens and renovating existing housing stock. The chapters explore the extent of energy poverty in each country and examine the drivers, while casting light on how policy-makers tackle the issue through a critical examination of the instruments implemented to help energy poor people.

This book will be of great interest to researchers in the fields of energy policy and comparative politics, to policy-makers in post-communist countries and EU institutions, and also to other relevant actors, such as companies and NGOs who focus on issues of energy poverty.

George Jiglau is Researcher and Lecturer in Political Science at the Babes-Bolyai University, Cluj-Napoca, Romania.

Anca Sinea is an Energy Policy Researcher at the Center for the Study of Democracy, Babes-Bolyai University, Cluj-Napoca, Romania.

Ute Dubois is an Associate Professor of Economics at ISG International Business School, Paris, France.

Philipp Biermann is an Economics and Social Science Researcher at University of Magdeburg, Germany.

Routledge Explorations in Energy Studies

The Politics of Energy Security
Critical Security Studies, New Materialism and Governmentality
Johannes Kester

Renewable Energy for the Arctic
New Perspectives
Edited by Gisele M. Arruda

Decarbonising Electricity Made Simple
Andrew F. Crossland

Wind and Solar Energy Transition in China
Marius Korsnes

Sustainable Energy Education in the Arctic
The Role of Higher Education
Gisele M. Arruda

Electricity and Energy Transition in Nigeria
Norbert Edomah

Renewable Energy Uptake in Urban Latin America
Sustainable Technology in Mexico and Brazil
Alexandra Mallett

Energy Cooperation in South Asia
Utilizing Natural Resources for Peace and Sustainable Development
Mirza Sadaqat Huda

Perspectives on Energy Poverty in Post-Communist Europe
Edited by George Jiglau, Anca Sinea, Ute Dubois, and Philipp Biermann

For more information about this series, please visit: www.routledge.com/Routledge-Explorations-in-Energy-Studies/book-series/REENS

Perspectives on Energy Poverty in Post-Communist Europe

Edited by George Jiglau, Anca Sinea, Ute Dubois, and Philipp Biermann

Routledge
Taylor & Francis Group
LONDON AND NEW YORK

earthscan
from Routledge

First published 2021
by Routledge
2 Park Square, Milton Park, Abingdon, Oxon OX14 4RN

and by Routledge
52 Vanderbilt Avenue, New York, NY 10017

Routledge is an imprint of the Taylor & Francis Group, an informa business

British Library Cataloguing-in-Publication Data
A catalogue record for this book is available from the British Library

Library of Congress Cataloging-in-Publication Data
A catalog record has been requested for this book

ISBN: 978-0-367-43052-8 (hbk)
ISBN: 978-1-003-00097-6 (ebk)

Typeset in Goudy
by Wearset Ltd, Boldon, Tyne and Wear

Contents

Figures

Tables

Contributors

Editors

Philipp Biermann is Researcher at the chair for Empirical Methods in Social Science, University of Magdeburg. He holds an MA in economics from the University of Göttingen (2012) and a PhD in economics from the University of Oldenburg (2017). His research interests are in behavioral economics, environmental and energy economics, distribution and justice, and applied quantitative econometrics. He is currently involved in the project Energy for Well-Being (Swiss Federal Office of Energy) and in the ENGAGER (COST Association) network.

Ute Dubois is Associate Professor of economics at ISG International Business School in Paris (France). Her main research topic is energy poverty with a focus on public policy analysis and qualitative social research. Ute's background is in energy economics and new institutional economics. In her doctoral research at University Paris Sud, she analyzed the governance of competitive reforms in the electricity sector, with a focus on Germany. In the past 12 years, her research has focused on the analysis of energy poverty policies in France and in Europe, on the question of identification of energy poor households, and on the relations between energy poverty and health.

George Jiglau is researcher and lecturer in political science at the Babes-Bolyai University in Cluj-Napoca, Romania, and president of the think tank Center for the Study of Democracy. He gained a PhD in 2013. His background interests lie in the broader sphere of comparative politics, with a focus on political parties, elections and electoral systems, and the development and behavior of political institutions, with a particular emphasis on local politics and its impact on quality of life. Starting from an interest in Europeanization and EU politics, he recently became increasingly interested in how the development of strong urban localities poses an additional challenge to traditional nation states, and in the impact the interplay between globalization and "glocalization" has on traditional institutional elements of political systems. Since 2016 he has become active on the topic of energy poverty, which he has integrated with his previously mentioned interests.

Anca Sinea is an energy policy researcher with expertise in the field of energy poverty. Her research work originates in the area of international relations as she holds an MA from the Diplomatic Academy of Vienna, where she centered on conflictual relations in the field of energy, and a PhD from the Babes-Bolyai University (BBU) in Cluj-Napoca with a focus on energy cooperation and multilateral diplomacy in the field of resources in Central and Eastern Europe. Over the years she has been involved in academic research, teaching, and consulting in the private sector and the policy-making process, in matters related to both energy poverty and security. Currently she is affiliated with the Center for the Study of Democracy think tank of the BBU and holds a board member position in the ENGAGER COST Action on energy poverty.

Contributors

Anna Zsófia Bajomi is a PhD candidate in urban planning, design, and policy at the Polytechnic University of Milan. Her research focuses on the application of EU policies supporting domestic energy efficiency and their effect on vulnerable households in Central and Eastern Europe, and especially Hungary. She holds a master's degree in social policy from the Eötvös Lóránd University of Budapest. She has work experience at local and international public bodies, research institutions, and NGOs. She has been engaged in different research projects related to energy poverty, such as the ENGAGER network.

Dušana Dokupilová works as a researcher at the Slovak Academy of Sciences, Centre of Social and Psychological Sciences, Institute for Forecasting. She works as a data scientist. Her work is dedicated to the environment, and is especially concerned with climate change. She has authored several studies connected with health, environment, and energy. Energy poverty is currently her main field of interest.

Nóra Feldmár holds a BSc in product design and an MSc in industrial ecology. She currently works at Habitat for Humanity Hungary as an energy poverty expert, which includes being one of the coordinators of the rEdistributor (Elosztó) project. Additionally, she focuses on technologies that are applicable, accessible, and can be sustainable for the most disadvantaged communities in Hungary and other Central and Eastern European countries. She has been engaged in practical projects related to rural energy poverty (especially within segregated communities) as well as conducting participatory research in the field.

Richard Filčák is Head of the Institute and Senior Researcher at the Slovak Academy of Sciences/CSPS-Institute for Forecasting. He has extensive experience as a researcher and development and social projects coordinator, working in Slovakia, Central Europe, West Balkan, Turkey, Ukraine and Georgia. Generally, his work and research interests are focused on environmental and social policy development and trends in the transition countries of Central and Eastern Europe – with particular attention on EU economic and social cohesion policies on industrial development and environmental trends. His main research interests involve energy poverty, poverty/environmental justice, access to resources (e.g., energy, water, safe environment), climate change, and implications for economic and social policies. He is the author of several books and professional peer-reviewed articles and research reports, and has worked as consultant and expert for the Slovak government, European Commission, UDP, UNEP, World Bank, and various NGOs.

Maria Jeliazkova is an associate professor and PhD in sociology. She is head of the Public Policies and Social Change department at IPS-BAS and teaches social policy at Sofia University and the University of National and World Economy. She works in the fields of anti-poverty strategies, employment and unemployment, and inequalities and welfare, and has authored over 70 publications.

Hedvika Koďousková is an energy expert with a special focus on social aspects of energy systems, energy policies in Asia, and Chinese energy policies. Currently, she is taking part in educational and research projects on energy poverty in the Czech Republic and Central and Eastern European countries, and on the central heating–energy poverty nexus. In her PhD, obtained in 2012, she analyzed the energy policies of China toward its neighbors. She has taken part in several governmental and academic research projects dealing with gas markets in general and Russian activities in gas and nuclear markets in particular and has published numerous articles and books on energy policies. In 2009 and 2011 she completed study visits in the UK (under the UK Energy Research Centre) and China (under the London School of Economics and Political Science/Beijing University). Previously, she worked as a team member and manager of a project focusing on students' education in energy policies co-financed by the European Social Fund and the Czech state budget.

Lea Kőszeghy holds a PhD in sociology from Eötvös Loránd University, Budapest (2011). Her specific research fields are housing poverty and housing policies, and the sociology of urban planning. She has worked in both the academic and the NGO sector. Since 2017 she is research fellow at the Institute for Sociology, Centre for Social Sciences, Hungarian Academy of Sciences Centre of Excellence. Between 2017 and 2019 she led research on energy poverty in Hungary in the research framework of societal challenges of energy use.

Veneta Krasteva works as associate professor in the Public Policies and Social Changes department of the Institute of Philosophy and Sociology, Bulgarian Academy of Sciences. She holds a PhD in sociology and carries out research on poverty and social exclusion, public policies reforms, and youth.

Lukáš Lehotský is an energy expert with a special focus on social opposition to coal mining, as well as the future of coal in the Czech Republic and the wider region. Also, he has dealt with issues of climate change, energy poverty, and energy efficiency. Before completing his PhD in 2018, he worked on a research project on energy efficiency in the Visegrad region and Ukraine in cooperation with the Slovak Foreign Policy Association in 2015–2016. Since 2017, he has worked primarily on research into local opposition to coal mining in the Czech Republic and has co-authored several articles on the topic. He also cooperated on several research projects outside the scope of his interest, in particular on gas market integration in the Visegrad region and on Russian activities in the gas and nuclear industries in Central and Eastern Europe.

Piotr Lewandowski is a labor economist and the President of the Board at the Institute for Structural Research (IBS), and a Research Fellow at the Institute for the Study of Labor (IZA) in Bonn, Germany. He was previously at the Warsaw School of Economics (2005–2010) and has collaborated with the Ministry of Labour and Social Policy and UNDP Poland. He is a cofounder of IBS and a project coordinator/principal investigator in national and international research projects in the fields of labor economics, demography, development, and energy economics (funded or commissioned by, for example, the National Science Centre in Poland, the World Bank, OECD, the European Commission, various ministries and central authorities in Poland). He is a co-coordinator of the Network for Jobs and Development, a global network group of research institutions from various regions of the world which build a vibrant community by providing a comprehensive understanding of the jobs agenda, including the effects of technological transformations around the world. His research is focused on the influence of technology on employment and labor market institutions as well as the impact of climate and energy policies on the labor market and living standards. He coordinated three research projects on energy poverty and energy efficiency in housing in Poland including cooperation with experts from leading centers in Europe (projects funded by the European Climate Foundation).

Douhomir Minev is a professor and doctor of sciences in sociology and also holds a PhD in economics. He works in the fields of economic and social development, social risks, and knowledge society. He has long-term experience in public policy review, analysis, and recommendations and has more than one hundred publications as a social scientist.

Lina Murauskaite (PhD) is a researcher at the Lithuanian Energy Institute, Laboratory of Renewable Energy and Energy Efficiency. Her field of research is renewable energy sources, energy efficiency, district heating, renewable energy communities, and energy poverty.

Jakub Sokołowski is an economist affiliated with the Institute of Structural Research and a lecturer in microeconomics at the University of Warsaw. He has been working in the energy sector for the past eight years. At the Institute for Structural Research he is mainly responsible for research on energy poverty. Together with the Institute's team, he has created a definition of energy poverty in Poland, and a multidimensional energy poverty indicator.

Joanna Stępień works as assistant professor in the Department of Socio-Economic Geography at the University of Gdansk, Poland. Her main research interests concern environmental and social determinants of urban health, healthy urban planning, environmental justice, and social equity. Before joining the University of Gdansk, she working for ten years for governmental institutions (both local and central) in Poland.

Ana Stojilovska is a PhD candidate in environmental sciences and policy at the Central European University, Budapest. She is studying the synergies between energy poverty, energy justice, and the type of heating in the developing and developed European context by using North Macedonia and Austria as empirical case studies. She is an active member of the COST-funded ENGAGER project. She was a Visiting Teaching Fellow at Eötvös Loránd University, Budapest, in 2018–2019, teaching energy justice and energy poverty. She has completed research fellowships at the Association for International Affairs, Prague, in 2012 and at the Autonomous University of Barcelona in 2019. She has fieldwork experience in several countries in Central and Eastern Europe. She has worked for Macedonian civil society as an energy researcher at Analytica Think Tank in Skopje and has been involved in project development, policy research, consultancy, and advocacy. She holds a master's degree in European studies from Europa-Kolleg Hamburg, University of Hamburg.

Tomislav Tkalec is the energy program coordinator of the Focus Association for Sustainable Development. His work is mainly focused on the topics of energy transition, energy efficiency, renewables, energy poverty, community energy projects, energy policies, and political ecology. He is experienced in working on the FP7, IEE, HORIZON2020 and LIFE projects, and is coordinator of the Sustainable Energy Working Group in the national network of environmental organizations, Plan B. His other tasks include advocacy and policy work, educational and awareness-raising activities, and communication with the media. His thesis, *Political Aspects of Decentralisation of Electricity Generation*, earned him a PhD in environment protection at the University of Ljubljana.

Damian Zelewski is a lawyer and urban geographer and PhD candidate in the Department of Socio-Economic Geography at the University of Gdansk. His main research interests concern urban planning, municipal social policies, revitalization, and legal aspects of spatial management.

Lidija Živčič is a senior expert at the Focus Association for Sustainable Development in Ljubljana, Slovenia. Her topics of work are sustainable development, climate, energy, energy poverty, transport, consumption, active citizenship, and degrowth. She has almost 20 years of experience in managing projects at the level of Slovenia and the EU. Currently, she is coordinating the H2020 project on energy poverty, EmpowerMed (2019–2023). After completing her undergraduate studies in economics in Ljubljana in 2000, she completed an MSc course in environmental science and policy at the Central European University in Budapest. In 2012, she earned a PhD at the biotechnical faculty, University of Ljubljana, with a dissertation on raising awareness on climate change in Slovenia.

Acknowledgments

The editors would like to thank Mithra Moezzi for her valuable contribution in preparing the manuscript.

Foreword

Stefan Bouzarovski

I am writing this in lockdown, in the spring of 2020, as the Covid-19 crisis rages across the world. There is no better or worse time for a contribution on this topic. People's confinement to domestic spaces due to Covid-19 preventive measures has foregrounded old and new socio-technical complexities associated with residential energy use. Many such challenges are associated with existing vulnerabilities embedded in the household domain, connected to energy-related income receipt and bill payment, mental health and well-being, as well as the incongruence between the fluidity of contemporary household arrangements and the fixity of residential enclosure. Had the first wave of the crisis hit the Global North in a period of more extreme temperatures, the consequences for vulnerable groups could have been much more grave.

Domestic energy vulnerabilities are particularly pronounced in the post-communist countries of Central and Eastern Europe (CEE), which in some instances contain the highest rates of energy poverty within the whole of the European Union. Energy poverty in this part of the world is not solely a question of household energy affordability, since many urban and rural centers lack infrastructural networks for the provision of energy services. As such, the region conforms to the broadest possible definition of energy poverty – the inability to secure socially and materially necessitated levels of energy services in the home (Thomson *et al.* 2019). In CEE, energy poverty is a systemic phenomenon that is deeply embedded in the regulation and management of the energy, housing, and social welfare sectors.

The reasons for the greater prevalence of energy poverty in post-socialist states are well known – they have been traced back to post-1989 rapid increases in energy prices, low household earnings, and a rise in general income poverty, as well the historically poor quality and energy efficiency of housing stock and relevant infrastructures. While most CEE states inherited a relatively similar set of legacies from the centrally planned economy, there has been significant divergence over the last 30 years, to the extent that some countries are in a better position relative to older members of the European Union, particularly those in southern Europe. Still, in most of

CEE, there are significant challenges around energy poverty, in research and policy terms alike.

The post-communist transformation has both shaped and been shaped by energy poverty. Here, I refer not only to the well-established effect of neoliberal economic restructuring, expressed through dynamics of utility privatization and price liberalization, as well as reductions in social welfare provision and housing restructuring. The crises of democracy and political representation that have engulfed some countries of the region have also affected social inequality and exclusion, impacting the ability of vulnerable households to access the support that they need, or to be recognized as being worthy of targeted policies in the first place. At the same time, rising rates of energy poverty have themselves influenced restructuring decisions taken during the post-socialist transformation, creating "lock-ins" that will configure some regions' and cities' socio-technical development for the foreseeable future (Bouzarovski *et al.* 2016; Ürge-Vorsatz and Tirado Herrero 2012). Energy poverty is creating new patterns of socio-spatial segregation that are just now beginning to be uncovered due to the relatively private and physically "hidden" nature of the problem (Bouzarovski and Thomson 2018; Grossmann 2019). It has become entangled in how an entire range of related inequalities are framed and experienced: from gender inequalities, to infrastructure, green space, gentrification, and air pollution (Bouzarovski *et al.* 2018; Frankowski 2020; Petrova and Simcock 2019; Tuvikene *et al.* 2019).

It is delightful to see a collection of essays that tackles all of these issues in such a comprehensive and thoughtful manner. Two decades ago, when I started working on energy poverty in CEE, expert interest in the topic was almost non-existent. There was only a relatively obscure and narrow debate on the income effects of energy prices, led principally by international financial institutions (Buzar 2007). Energy-poverty-associated challenges in some countries of the region had received almost no scientific attention – a situation mirrored in the policy arena. But the chapters in this volume testify to how far we have all come in creating an informed polity of academic research, advocacy work, and on-the-ground engagement. In its entirety, the stakeholder community embedded in such efforts is helping ensure that energy poverty concerns are not forgotten as we develop concerted initiatives to move toward a low-carbon future in the European Union and beyond.

The chapters that follow stem from professional collaborations and discussions that have taken place under the auspices of an Action funded by the European Co-operation in Science and Technology, titled "European Energy Poverty: Agenda Co-Creation and Knowledge Innovation" (ENGAGER 2017–2021). As such, they testify to the incredible importance of international knowledge exchange and scientific cooperation at a time when many political forces are advocating for a retreat toward much more limited spatial and temporal horizons. I congratulate the editors and contributors for providing us with the only comprehensive review of energy poverty in CEE to date. It will set research and policy agendas for years to come.

References

Bouzarovski S and Thomson H 2018 Energy vulnerability in the grain of the city: Toward neighborhood typologies of material deprivation *Annals of the American Association of Geographers* 108 695–717.

Bouzarovski S, Frankowski J and Tirado Herrero S 2018 Low-carbon gentrification: When climate change encounters residential displacement *International Journal of Urban and Regional Research* 42 845–863.

Bouzarovski S, Sỳkora L and Matoušek R 2016 Locked-in post-socialism: Rolling path dependencies in Liberec's district heating system *Eurasian Geography and Economics* 57 4–5.

Buzar S 2007 *Energy Poverty in Eastern Europe: Hidden Geographies of Deprivation* Aldershot, Ashgate.

Frankowski J 2020 Attention: smog alert! Citizen engagement for clean air and its consequences for fuel poverty in Poland *Energy and Buildings* doi: 10.1016/j.enbuild.2019.109525109525.

Grossmann K 2019 Energy efficiency for whom? A conceptual view on retrofitting, residential segregation and the housing market *Sociologia Urbana e Rurale* FrancoAngeli Editore 2019 78–95.

Petrova S and Simcock N 2019 Gender and energy: Domestic inequities reconsidered *Social and Cultural Geography* doi: 10.1080/14649365.2019.1645200.

Thomson H, Simcock N, Bouzarovski S and Petrova S 2019 Energy poverty and indoor cooling: An overlooked issue in Europe *Energy and Buildings* 196 21–29.

Tuvikene T, Sgibnev W and Neugebauer C S 2019 Introduction: Linking post-socialist and urban infrastructures, in Tuvikene T, W Sgibnev and C S Neugebauer eds *Post-Socialist Urban Infrastructures* London and New York, Routledge 1–20.

Ürge-Vorsatz D and Tirado Herrero S 2012 Building synergies between climate change mitigation and energy poverty alleviation *Energy Policy* 49 83–90.

1 Introduction

Energy poverty and its drivers in post-communist Europe – the visible, the measurable, and the hidden

Ute Dubois

The slow emergence of energy poverty on policy agendas in post-communist Europe

In comparison to energy poverty levels in other European countries, those in countries of post-communist Europe are high (Bouzarovski and Thomson, 2019; Bollino and Botti, 2017; Dubois and Meier, 2016; Thomson and Snell, 2013). However, characterizing energy poverty in the context of post-communist Europe is not an easy task. Looking at European statistics, it appears that there is not one type of energy poverty but rather various forms, which do not have the same salience in all countries.

As shown by data from the European Energy Poverty Observatory,[1] some post-communist countries rank highest on two main energy poverty indicators: "inability to keep home adequately warm" and "arrears on utility bills." On the "inability to keep home adequately warm" indicator, Bulgaria and Lithuania are the two most affected countries, with 33.7 percent and 27.9 percent respectively of households affected in 2018, compared to the EU average of 7.3 percent. On the "arrears on utility bills" indicator, the group of the five most affected countries includes four post-communist countries: Greece – the most affected country in Europe (35.6 percent) – is followed by Bulgaria (30.1 percent of households), Serbia (28.4 percent), Croatia (17.5 percent) and Romania (14.4 percent).

Despite the sometimes-severe difficulties of households, no single method allows satisfactory characterization of energy poverty throughout the region. On many indicators produced by the European Energy Poverty Observatory, energy poverty symptoms do not have similar intensities throughout the region, and energy poverty occurs in a less visible way. For example, in Romania, one of the lowest-income countries in Europe, 51.4 percent of the households in the first income decile had low absolute energy expenses in 2015, which suggests that the poorest households are either massively rationing their energy use or that they are obliged to rely on low-cost or free fuels (such as wood collection or burning waste). The same problem appears in Poland, where 44.4 percent of households in the first income decile have low absolute energy expenses. Looking at the indicators measuring whether dwellings are comfortably warm in winter and comfortably cool in summer, Bulgaria ranks second on "dwelling not comfortably warm in winter," with 42.3 percent of households affected in 2012, and ranks first on "dwelling not comfortably cool in summer," with 52.4 percent of households affected.

However, energy poverty has been entering the policy agendas of countries of the region only slowly and the causes of energy poverty are sometimes not well recognized by policy-makers. In some countries, such as Poland, the first analyses on the topic were developed several years ago (Stępniak and Tomaszewska, 2013). But most countries are still in the initial stages of the recognition of energy poverty. While some countries, for example the Czech Republic, recognize that energy poverty is a multidimensional problem,[2] others, such as Slovakia, consider that energy poverty is mainly a problem of poverty.[3] In the recent report of the European Observatory of Energy Poverty, which includes a more general overview of the approaches of different countries on the issue of energy poverty (Bouzarovski and Thomson, 2019), we learn for example that Lithuania and Hungary, despite being significantly affected by energy poverty, "consider that the measures already in place (such as price regulation in Hungary), are sufficient to address the phenomenon" (EPOV, 2019). Bulgaria – one of the countries with the highest energy poverty levels – focuses on creating "the right conditions for the protection of consumers in the framework of an open market." In Estonia, social welfare seems to be the principal approach for tackling energy poverty. Poland mentions the role of "energy efficiency housing upgrading strategies for energy poverty" to reduce CO_2 emissions and improve air quality. And Romania "acknowledges the prevalence of energy poverty and mentions several strategies to tackle it," including renovation programs addressing explicitly energy poor people. Slovenia, "unlike many other EU countries … focuses on improving energy efficiency and retrofitting the dwellings occupied by the most vulnerable." And Croatia has announced an integrated program for elimination of energy poverty (EPOV, 2019).

In many countries of the region, the policy measures and level of support given to energy poor households are only limited, leaving households exposed to cold homes, increasing energy burdens, and debts with energy suppliers.

This weak presence of the topic on national policy agendas also translates to research efforts dedicated to energy poverty analysis that are less developed in post-communist Europe than in other countries. As noted by Gangale and Mengolini (2019), participation of researchers and institutions of the region in EU research and innovation projects is still low, despite high levels of energy poverty in some countries. This situation is perhaps explained by the fact that past research projects have focused to a large extent on social housing, which represents only a small share of homes in eastern Europe (Pittini et al., 2017; Hegedüs, 2011).

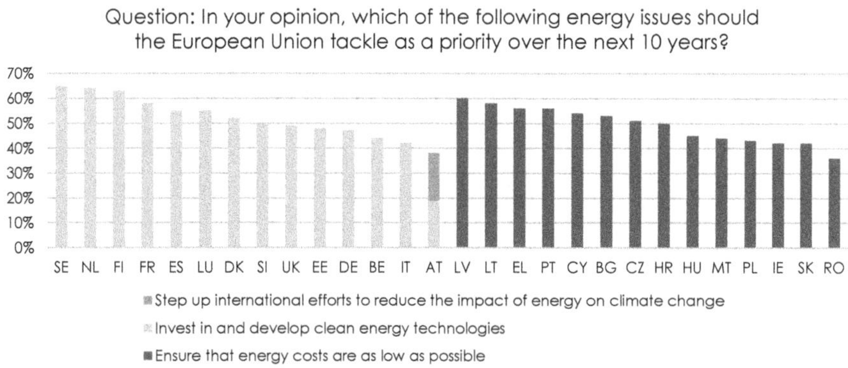

Figure 1.1 Expectations of European citizens regarding future European energy policy.
Source: European Union (2019).

Still, in a context where energy policies are increasingly defined at the European level, households of the region have clear expectations regarding the goals of future European policies. As shown in Figure 1.1, drawn from a recent Eurobarometer survey, respondents from the eastern part of Europe expect the European Union to prioritize improving the affordability of energy.

In this chapter, we contend that many problems related to energy poverty do not come to the forefront of national debates in the region because some of their aspects are invisible or scarcely visible to policy-makers. The chapter has been developed following exchanges among energy poverty experts that have taken place since the launch of ENGAGER COST Action in 2017. In particular, the meetings to prepare the present book have helped highlight the factors driving energy poverty in the context of post-communist Europe.

Our main contribution is twofold. On the one hand, we discuss these visibility problems and their implications. On the other hand, we use the literature and exchanges with energy experts from the region to highlight certain specificities of energy poverty in post-communist Europe. This chapter is organized into three parts. The first section discusses the lack of visibility of energy poverty symptoms for policy-makers, which implies that efforts need to be undertaken to acknowledge the issue. The second section analyses the variety of energy poverty drivers and context factors and shows that many determinants of energy poverty are actually influenced not by individual characteristics of households but by factors which are part of the larger context, which can be national or regional. Finally, the third section discusses some of these context factors which might be particularly relevant in the case of post-communist Europe.

Energy poverty symptoms are not all equally visible

Energy poverty, as a particular form of material deprivation, is present in most societies (Bouzarovski and Petrova, 2015; Birol, 2007). Under these circumstances, how can we understand that energy poverty is still lacking recognition in policy and public debates in so many countries? One reason might be the fact that energy poverty is multidimensional (Rademaekers et al., 2016). The evolution of measurement methods (Sareen et al., 2020; Thomson et al., 2017; Hills, 2012) has allowed the dimensions of energy poverty to be captured more accurately. However, the concept of energy poverty is still hard to grasp for policy-makers who have little experience with the issue.

In this section, we will not go into the details of energy poverty measurement methods. Rather, the main argument developed here is that energy poverty is often hard to grasp – for the general public and partially for policy-makers – because some of its aspects remain hidden. Few of the many symptoms of energy poverty are directly visible to stakeholders; the other symptoms need to be measured, or evidence on their existence needs to be collected to make them visible.

Some of the most visible symptoms of energy poverty started being discussed several decades ago. For example, the existence of utility debts has been reported by suppliers for many years, especially since the liberalization of network industries (Graham, 2006. In some countries, such as France, utility debts were not initially labeled as energy poverty. However, suppliers recording increasing levels of debts was a signal that some energy consumers were faced with affordability problems. The issue of utility debts is important for suppliers especially in contexts where energy pricing is modified to better reflect the actual costs of energy supply, as has been observed in many post-communist countries. But when energy prices are widely subsidized or when cross-subsidies between groups of consumers – for example between industry and domestic consumers – limit the impact on the population of price increases, the potential problem of energy affordability may be less visible to policy-makers because these subsidies artificially reduce the scale of energy affordability difficulties of households.

Another visible symptom of energy poverty is household demands for financial assistance when they are unable to pay their energy bills. These symptoms become visible mainly when some assistance schemes are available. Where assistance schemes do not exist, households might be cut off by suppliers, or they drastically ration their energy consumption, which is a more invisible form of energy poverty. Finally, one element that has gained visibility, even among the general population, are accidents such as fires caused by households who have been cut off from their electricity supply. In several countries – for example Germany, Spain, and Bulgaria – such events have created moments when difficulties related to energy poverty have become visible and are publicly debated.

But less visible energy poverty symptoms require the production of information and elaboration of data, either by compiling existing statistics or by setting up dedicated surveys. For example, to estimate excess winter mortality or the share of households' budgets dedicated to energy, statistical data are used to build estimates of energy poverty symptoms. Other surveys, for example the EU Survey on Incomes and Living Conditions, specifically collect data on elements of energy poverty, such as the inability of households to keep their home adequately warm.

Finally, there is a group of energy poverty symptoms that is often implicitly known, but which is nonetheless hard to identify or measure, either because these symptoms are hard to quantify – for example health impacts – or because they occur in the private sphere of households that are often reluctant to speak about their difficulties. Frequently, households find it easier to talk about their low income than about restrictions they impose on themselves, or about shame, stigma, and social isolation resulting from energy poverty.

The different degrees of visibility of energy poverty symptoms are summarized in Figure 1.2. This figure highlights that most energy poverty symptoms need either to be measured – which requires setting up measurement systems and rules – or to be made visible, through in-depth examination of the relations between energy poverty, health and well-being, behaviors, lived experiences, and perceptions of the quality of life of affected households.

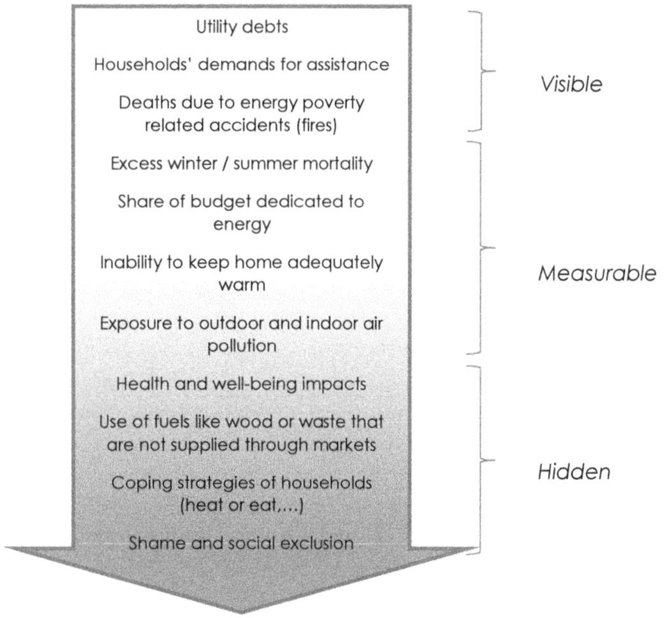

Figure 1.2 Energy poverty symptoms and their visibility.

Source: Author's analysis.

In the countries of post-communist Europe, the problem of energy poverty might remain hidden if policy-makers focus mainly on its visible symptoms, like utility debts and demands for financial assistance of households. These symptoms are frequently interpreted as a result of poverty in general, or of a lack of affordable energy. In certain cases, policy-makers respond to people's demands for more-affordable energy by subsidizing energy prices, or by imposing price reductions through political decisions.[4] Interpreting energy poverty only as a problem of low incomes or of energy affordability has two consequences. First, in terms of remedies, it would imply measures of income support or price reductions for a large share of the population, which is neither feasible nor desirable. Second, interpreting energy poverty as a result of general poverty comes at the risk of missing an important part of the problem of energy poverty, which is lacking energy efficiency of homes and equipment. When Boardman (2010) recommends that "the poorest people should have the most energy-efficient homes," she insists on that distinction between energy poverty and poverty.

Energy poverty measurements in several countries have shown that there is only a partial overlap between the group of poor households and the group of energy poor households. This confirms that interpreting the visible symptoms of energy poverty simply as a problem of poverty is inadequate. On the other hand, if energy poverty is viewed as a problem of poor housing conditions and of material deprivation, then it becomes clear that the exact nature of the difficulties of households needs to be investigated further. Uncovering the less visible aspects of energy poverty requires for example assessing the extent of difficulties of households in keeping their home warm. It can also include assessments of the lived mechanisms of energy poverty. As shown by an analysis of the European Energy Poverty Observatory (Bouzarovksi and Thomson, 2019), not all countries of the region have started measuring energy poverty even with very simple measurement tools, leaving their national and local decision-makers unaware of the extent of the problem.

Some hidden aspects of energy poverty have been mentioned in the discussions among energy poverty experts which took place at various ENGAGER COST meetings and during the work on this book. In the region, energy poor households often adopt rationing behaviors, like heating only one room, or being obliged to use electric heaters to complement other forms of heating. One extreme case occurs when people need to choose between heating or eating (eating less or lower quality food). When heat is supplied through district heating, households need to cope with supply interruptions of the heating system[5] and, more generally, with the inefficiencies of district heating, like the absence of possibilities to choose the temperature inside homes, resulting in both over-heating and under-heating of homes, the interruption of heating during the night, which obliges people to use electric heaters as a backup, or the heating cost allocation rules, which sometimes depend at least partially on the size of the dwelling or on the heated volume.

Other hidden aspects include the necessity for certain households to heat their homes with wood or with waste, for example in rural areas. Some experts mentioned that burning solid fuels can give people a feeling of being in control of their heating situation. Sometimes households do not pay for the energy they use (wood theft, illegal wood cutting) or they resort to the gray economy or to informal systems for the provision of energy. Burning solid fuels can result in the development of smoke inside the home and resulting health problems for households. But when these inefficient heating systems are regulated, for example to limit air pollution, this can result in further increases of energy poverty, as some households cannot afford the expense of other commercial fuels or to replace their heating systems.

Among the other, more informal, elements of energy poverty that are hard to grasp are energy consumption cultures and habits such as those inherited from previous generations, where energy was cheap and was not something individual households needed to care about. Another aspect experts mentioned is the lack of trust in the government, for example when renovation programs are offered to energy poor households.

These invisible symptoms of energy poverty and their impacts can be hard to grasp for policy-makers. In general, knowledge on these aspects needs to be formally elaborated. One study showing some invisible aspects of energy poverty is the analysis of Druică et al. (2019) on a set of countries of Central and Eastern Europe.[6] They show that energy poverty does not only result in health problems, but also leads to "low levels of self-esteem and consequently to diminished satisfaction with life," what the authors call "meaning-making – the manner in which individuals translate into their own understanding both their personal and different social contexts." One lesson of this work is to make explicit the relation between energy poverty and people's life satisfaction. However, the authors note that improvements in life satisfaction may not be a sufficient motivation for policy-makers to act, as people's dissatisfaction is something "that neither translates into salient costs, as explicit health issues do … nor triggers extreme interest for public decision makers."

Going beyond incomes, energy efficiency and prices: individual energy poverty drivers and larger context factors

The main drivers of energy poverty as identified by Boardman (2010) include low incomes, high energy prices, and poor energy efficiency of homes. In the past decade, there has been abundant literature exploring the role of these drivers, both in case studies and in comparative analyses (Pye et al., 2015). Table 1.1 lists the drivers and context factors contributing to energy poverty

Table 1.1 Energy poverty drivers and context factors

Low incomes	*Poor energy efficiency of homes and equipment*	*High energy prices and poor energy supply conditions*
Individual drivers	**Individual drivers**	**Individual drivers**
– Material deprivation – Age, gender – Household composition – Employment status, vulnerable employment – Level of education – Health status – Ethnicity	– Financial capacity to invest – Heating equipment (age and quality)	– Ability and financial capacity to choose sources of energy used for heating
Both individual and contextual	**Both individual and contextual**	**Both individual and contextual**
– Feeling poor – Past and expected income evolution – Trust in institutions – Dependency on remittances – Importance of the shadow economy – Population dynamics (emigration, ageing)	– Age and type of buildings (for example, panel buildings) – Energy management inside buildings (case of district heating) – Individual/collective housing – Home ownership – Urban/peri-urban/rural – General quality and condition of building (renovated or not) – Exposure to cold or warm temperatures in different locations – Trust in renovation policies and programs – Trust in neighbors (for the renovation of co-owned buildings)	– Interdependencies between energy consumers, in cities and in buildings (use of district heating)
Context factors	**Context factors**	**Context factors**
– Income inequalities (between income groups, inequalities related to age, gender, education level) – Social policy – Economic crises – Regions in economic decline – Post-war country – Quality of institutions	– Ease of renovating buildings – Regulation of building quality – Regulation of heating equipment – Past and present refurbishment policies – Population dynamics (local and national) including emigration influencing desirability of renovations locally – Local climate	– Existence and availability of various and affordable energy sources – Access to networks supplying affordable energy sources – Quality of energy supply infrastructures – Right to opt out of certain types of energy supply (for example district heating) – Protections of vulnerable energy customers – Energy price regulation – Subsidized household energy prices – Energy sector reforms (competition) and quality of regulation – Share of households vs business or industry electricity customers in the country – Regulation of the use of coal or waste to limit air pollution – Evolution of local energy markets (for example, wood) in terms of price and availability – Energy security and independence considerations of the country

Source: author's analysis.

that have been raised in these discussions, as well as in the literature (Primc et al., 2019b; Bollino and Botti, 2017; Rademaekers et al., 2016). We classify these factors into the broad categories of incomes, energy efficiency of homes, and energy prices and supply conditions. We also differentiate between those elements which are individual, at the "micro-level," and those that can be considered as contextual, which instead concern some "macro-level" factors (Bollino and Botti, 2017; Dubois and Meier, 2016; Rademaekers et al., 2016).[7] Among the contextual factors, we include those factors that cannot be influenced by individual households. Between the individual drivers and the contextual factors, we add a hybrid category, which includes factors that are partly individual and partly contextual. These elements can be to some extent determined individually, but they also have a collective, or contextual, dimension. The ambition of this list is not to cover all possible factors that might play a role, but rather to highlight the importance of drivers which are not exclusively individual.

This list of drivers has several implications. First, it helps thinking about factors that are embedded in the economic, political, and demographic structures and resource endowments of countries and that contribute to energy poverty. Recognizing this embeddedness of energy poverty drivers is important because it implies that remedies to energy poverty may require the deeper roots of the problem to be addressed.

Second, the table suggests that, for all three types of drivers – incomes, energy efficiency, and prices and supply conditions – there are numerous factors of a larger scale beyond solely the individual. For energy poverty researchers, this implies that these larger-scale factors deserve further analysis. Often, international organizations measure such factors at the country level. For researchers, this means that the impact of these determinants on energy poverty levels and on the depth of energy poverty in different countries can be tested, as has already been done by Primc et al. (2019a and 2019b) and Bollino and Botti (2017).

One direct use of our list of drivers and context factors is the possibility to examine, for each country, the factors with which countries have the most severe difficulties. This allows national energy poverty contexts to be described and those factors which are most characteristic of each country to be highlighted. In a comparative perspective, this allows the specificities of countries with respect to some of these factors to be portrayed. Finally, at the regional level, extending the list of drivers and context factors could be helpful to identify those elements which are distinctive of post-communist Europe.

Distinctive characteristics of post-communist Europe regarding energy poverty drivers and context factors

Focusing on post-communist Europe, Bouzarovski et al. (2012) consider that energy poverty in the region is shaped by "indirect energy price subsidies, reliance on polluting sources of energy, state interference and ownership of energy enterprises as well as inefficient housing stocks and heating systems." Since the launch of ENGAGER COST Action in 2017, the drivers of energy poverty have been discussed on several occasions. The meetings to prepare the present book were helpful to highlight those factors which contribute to energy poverty in the context of post-communist Europe and that we develop here.

Spatial disparities in terms of energy poverty prevalence

The literature on post-communist Europe has, since the beginning, highlighted the existence of spatial differences regarding energy poverty. In his analysis on (North) Macedonia and the Czech Republic, Buzar (2007) analyzed spatial patterns of energy poverty. He notes that these patterns are particularly influenced by the availability of and access to various energy sources. In North Macedonia, the increase in energy prices in the 1990s resulted in an expansion of the number of energy poor households especially in areas where fuelwood was not available and where district heating was not developed. In these areas, households were obliged to heat their homes with electricity. In comparison, in the Czech Republic, energy poverty appeared to be more linked with low incomes and more concentrated geographically in the poorest areas.

In their analysis of regional energy poverty patterns in Poland, Lis et al. (2016) show that significant differences between regions exist, both in the dimension of energy affordability (measured by a low-income high-cost indicator) and in the lack of thermal comfort, measured by a self-reported indicator. Affordability is significantly related to income and to the area of residence, with high levels in rural areas, whereas lack of thermal comfort is mostly observed in cities, which leads to the conclusion that thermal retrofits should be prioritized in cities.

Bouzarovski and Tirado Herrero (2017), in their analysis of regional energy poverty in Poland, the Czech Republic and Hungary, note a "tendency for domestic energy deprivation to be concentrated in rural and peripheral regions with poor-quality housing and decreased access to affordable fuels." The de-industrialization of certain regions and regionally unequal economic transition paths are also mentioned as important aspects shaping regional energy poverty disparities (Bouzarovski and Tirado Herrero, 2017).

At a European level, the heterogeneity of energy poverty factors across regions had already been noted by Velte et al. (2010), who showed that some elements such as climate, energy costs, unemployment, or age-dependency ratio are sometimes more similar between regions of different countries rather than within a given country. This observation leads to the conclusion that some regional specificities might be hidden when looking at the problem only at a national level. Therefore, in a recent ENGAGER

COST policy brief, we suggested the development of energy poverty analyses at the regional level in order to capture such local specificities more precisely (Dubois et al., 2018).

The discussions among energy poverty experts from the region have also highlighted the importance of spatial disparities in terms of energy poverty. While in certain countries, such as Hungary, rural areas seem to be most exposed to energy poverty, in other countries, such as North Macedonia, peri-urban energy poverty seems to be more severe. Migration – inside countries and between countries – seems to influence spatial energy patterns as well: in certain areas, the decline in population leaves many homes empty or occupied by pensioners, reducing incentives to invest in thermal retrofits. This is the case for example in Lithuania. In addition, the lack of a local workforce also contributes to reducing the number of renovations. In Slovakia, regional energy poverty patterns are influenced by regional economic and industrial dynamics. Finally, in Poland, energy poverty seems to follow some regional patterns in relation to coal production. On the one hand, policy debates on energy poverty in Poland are focused on local air pollution caused by the combustion of coal. On the other hand, the economic transition of the coal-producing regions away from coal results in lower incomes and higher levels of energy poverty in certain areas. These examples of determinants of spatial energy patterns show the variety of causal mechanisms that can play a role in energy poverty as well as their sometimes hard to measure aspect.

Poor targeting and segregation between social groups, contributing to increased energy poverty among certain populations and to situations of deep energy poverty

The targeting of energy poverty measures, when they exist, can be considered imperfect for different reasons. One reason is the almost exclusive focus on low-income households. Another reason is poor delivery of measures. Frequently, public policies addressing energy poverty focus on low-income households. However, as shown for example in the case of Poland (Rutkowski et al., 2018), there is an imperfect overlap between households on low incomes and energy poor households. In addition, it has been observed, for example in Romania, that many households who are in the target group do not actually benefit from assistance schemes. Murafa et al. (2017) note that in Romania less than 30 percent of eligible households actually receive heating allowances, concluding that "over half of the administrative effort of authorities with granting benefits goes to the households that are in the higher income" categories.

The discrepancy between social assistance and heating allowances on the one hand, and energy poverty estimations on the other hand is also highlighted by Lenz and Grgurev (2017). In their comparison of Bulgaria, Croatia, and Romania, they show that the number of beneficiaries of assistance schemes is generally below energy poverty estimates. Bulgaria is the country where the gap is the highest between beneficiaries of heating allowances and households unable to keep their home adequately warm, with 10 percent of households being beneficiaries of heating allowances in 2013 and 45 percent of households unable to keep their home adequately warm. In Croatia, 2.9 percent of households received heating allowances in 2013, while 10 percent of households were unable to keep their homes adequately warm and the share of energy in households' expenditures exceeded 10 percent for households in the income deciles 1 to 6. In Romania, the proportion of households benefiting from heating allowances and the share of households unable to keep their home adequately warm were similar – 14 percent of households – while the share of energy in households' expenditures exceeded 10 percent for households in the income deciles 5 to 9 (Lenz and Grgurev, 2017). This is consistent with the EPOV data on the low absolute energy expenditure indicator (M/2), which shows that Romania is one country of the region where many households are affected by this form of energy poverty. Lenz and Grgurev (2017) note that energy poverty in the three countries must be viewed in a larger context, where liberalization of energy markets and the transition from subsidized household energy prices to cost-reflective prices have resulted in "restrictions of comfort, inadequate heating … and high income share spent on energy bills."

Another element, one gaining increased attention, is the existence of certain energy poor households that are not traditionally visible, or who are not included in assistance schemes. In their European comparison, Bollino and Botti (2017) identify some energy poor groups that are traditionally misrepresented, such as private renters and migrants.

In post-communist Europe, the Roma population suffers from multiple forms of exclusion that result, among other problems, in high energy poverty levels among this population. The Roma communities suffer heavily from housing exclusion. Not only do they often live in segregated areas, which are mostly rural, they are also often excluded from labor market participation, live in poor housing conditions (Berescu et al., 2012), and are victims of various forms of "environmental racism" (Heidegger and Wiese, 2020). Many Roma lack access to essential services, like access to safe drinking water or sewerage (Kósa et al., 2011). The poor housing conditions in which they live result in situations of extreme energy poverty (Teschner et al., 2020) where informal housing, lack of documents – and thus the inability to apply for assistance schemes – and "gray" incomes all contribute to a certain form of invisibility of this population with regards to energy poverty policies. The Roma population represents between 8 and 10 percent of the total population in Slovakia, Romania, Hungary, and Bulgaria (Heidegger and Wiese, 2020). However, estimating this population is a difficult task. As shown by Kósa et al. (2011), the number of Roma living in colonies is often higher than the population measured by census data, which therefore "cannot be used for policy design aiming at those Roma who are in greatest need of help."

The discussions with energy poverty experts from the region have highlighted the importance of energy poverty among these special populations in Hungary, Slovakia, the Czech Republic, and Romania. These populations live in substandard housing in segregated territories and are extremely energy poor. Experts mentioned the existence of "ethnicized debates" in the media regarding the debts of the Roma, resulting in the Roma being considered as "undeserving poor."

Historical lack of investment in housing and in infrastructures

The state of the housing stock and the conditions for investing in improvements have certain distinctive features in post-communist Europe. In many countries, a high share of the current housing stock was built during the communist era. The panel buildings in urban areas are the most visible part of the post-communist heritage in terms of buildings. They are generally prefabricated constructions made with cheap materials. Having been built in a period of cheap/highly subsidized energy, they are also poorly insulated. Their heating systems, based on district heating, are generally considered inefficient. In the 1990s, these buildings were massively privatized. Therefore, today, many of them are owner occupied. The new homeowners, who are mostly the previous tenants of these flats, were generally unable to realize the investments that would have been necessary to upgrade these buildings. Moreover, the privatization of flats was not accompanied by the creation of "adequate regulatory frameworks to manage and maintain these buildings" (Kakalejclkova, 2017). The consequence has been an under-investment in buildings which has lasted long after the end of the communist regimes.

In Romania, the building stock inherited from the communist era also includes smaller homes constructed between 1945 and 1969, located in small villages and towns. These buildings, too, are characterized by poor building quality "due to lack of subsidies, shortages of durable construction materials and lack of infrastructure" (Soaita, 2014). These homes, together with the panel buildings, still represent a large part of the housing stock. As a consequence, Romania displays one of the poorest housing performances, with "low habitable space availability and high overcrowding levels" (Soaita, 2014).

In Bulgaria, Peneva (2014) estimated that 20 percent of homes were panel buildings in 2005, most of them needing renewal. In the majority of these buildings, district heating was the main heating source. However, the removal of subsidies for district heating prices together with increasing poverty levels resulted in many users disconnecting from the district heating network to use electric heating at the end of the 1990s. This situation makes households dependent on electricity prices. These are among the lowest energy prices in Europe, but, compared to the purchasing power of households, they remain high.

The situation in southeast Europe also fits this general description of poor energy efficiency of buildings combined with inefficient heating systems and an overall lack of maintenance of buildings and equipment (Robic et al., 2016):

> In some cases, the buildings are in such a deteriorated state that it is doubtful whether full energy efficiency retrofitting would be possible even if they would have the opportunity to do so. Many buildings are of substandard construction quality, and it would not be cost-effective to undergo refurbishment, hence re-settlement programs would be needed.
>
> (Robic et al., 2016)

The discussions among energy poverty experts from the region fit this general description. In Romania, the general infrastructure is considered old and requires massive investment. At the same time, a large share of the population is not reached by gas networks or certain other infrastructures. In many cities, district heating is highly subsidized, inefficient, and sometimes subject to supply cuts. In North Macedonia, the situation of district heating is similar to that in Bulgaria, with disconnections of households from the district heating networks. In Hungary and the Czech Republic, the withdrawal of the state from the housing sector and the poor quality of buildings were mentioned as important factors contributing to energy poverty. While these elements are well known in the case of post-communist countries, it is nevertheless important to include them in an analysis of the specificities of energy poverty in post-communist Europe. These specificities substantially shape the capacity of these countries to remedy energy poverty. This capacity is limited in the short term, due to the huge investment that would be required to massively refurbish the inefficient housing stock in the region.

Energy independence considerations of countries

In terms of energy markets, in past decades, the countries of the region had to switch from supply systems where energy was abundant and cheap to systems where energy supply is mostly based on market prices. This generally resulted in increases of energy prices for households, due to the removal of subsidies. At the same time, it seems that people are still sometimes lacking "the understanding that energy use needs to be managed properly, leading to often wasteful use of energy" (Robic et al., 2016).

In the discussions with energy poverty experts of the region, the topic of energy independence of countries was frequently raised in relation to energy poverty. Seeking greater independence from energy imports, policy-makers often tend to favor locally available energy sources, which can have effects in terms of energy poverty. Countries with local fossil fuel resources use these energy sources in priority. This is the case in Romania with natural gas, and in Poland and the Czech Republic with coal. One country with no fossil fuel resources is Lithuania. After the closure of a nuclear power plant there at the end of 2009, the country became heavily dependent on fossil fuel imports. The construction of a liquefied natural gas terminal has allowed the country to be less dependent on Russian gas imports. However, this investment came at a high cost for energy consumers through higher utility prices. Lithuania also significantly developed its share of renewable energy, and in particular the combustion of wood.

Another element mentioned in expert discussions was government intervention to prevent price increases in certain countries. For example, in Poland, parliament passed a law in December 2018 to freeze electricity prices. In Hungary, electricity prices increased between 2008 and 2012 and Hungary became the country with the highest household electricity prices of the region in terms of purchasing power standards. The subsequent nationalization of the electricity supply and the control of prices by the government allowed significant reductions in households' electricity bills.

The persistence of inefficient energy subsidy systems

More generally, several post-communist countries are still locked in policies where some energy sources are subsidized (Table 1.2). Such subsidies do not primarily consist of direct, or pre-tax, energy subsidies. They consist more frequently of inefficient levels of taxation of energy sources.[8] These can have negative social and environmental effects including, among others: local air pollution and subsequent health problems; higher greenhouse gas emissions; the discouragement of investments in energy efficiency, clean energy sources, and infrastructures; and the persistence of the dependence of countries on imported energy sources (Coady et al., 2015). Moreover, these energy subsidies are generally regressive: they mainly benefit the wealthiest households, at the expense of poorer households, who would rather benefit from a system of targeted subsidies.

Table 1.2 shows that in post-communist Europe the shares of gross domestic product dedicated to energy subsidies are generally higher than in the western part of Europe. In the countries with the highest energy subsidies, coal subsidies represent a high share of total subsidies, encouraging higher levels of local air pollution and impacts in terms of global warming. These subsidies have a hidden aspect, as their distributional impacts (in the form of lower energy costs) and their costs (local air pollution and health) are not visible unless they are calculated, for example by the International Monetary Fund.

Table 1.2 Energy subsidies in European countries in 2017

	Post-tax subsidies by component as a percentage of GDP											
	by product				*by component*							
	Petroleum	*Coal*	*Natural Gas*	*Electricity*	*Pre-tax subsidies*	*Global warming*	*Local air pollution*	*Congestion*	*Accidents*	*Road damage*	*Foregone consumption tax revenue*	*Total post-tax subsidies*
Advanced economies (Western Europe)												
Austria	0.61	0.23	0.06	–	0.01	0.41	0.23	0.28	0.03	0.02	(0.09)	0.89
Belgium	0.73	0.23	0.42	–	–	0.47	0.21	0.23	0.01	0.02	0.44	1.38
Cyprus	0.25	0.01	0.45	–	–	0.38	0.08	–	–	–	0.25	0.71
Denmark	1.84	0.15	0.15	–	–	0.32	0.17	1.17	0.04	0.03	0.40	2.13
Finland	0.67	0.36	0.14	–	0.00	0.39	0.30	0.15	0.01	0.01	0.31	1.17
France	1.02	0.15	0.21	–	0.00	0.31	0.32	0.41	0.03	0.03	0.28	1.39
Germany	0.88	1.18	0.31	–	0.02	0.69	1.01	0.21	0.02	0.01	0.40	2.37
Greece	0.37	1.62	0.18	–	0.00	0.74	1.13	0.03	0.01	0.00	0.26	2.17
Ireland	1.08	0.17	0.18	–	–	0.31	0.19	0.58	0.04	0.01	0.30	1.43
Italy	0.26	0.26	0.47	–	0.00	0.45	0.25	0.07	0.02	0.01	0.19	0.99
Luxembourg	3.95	0.10	0.21	–	–	0.53	0.46	2.79	0.04	0.00	0.42	4.26
Malta	0.52	1.30	0.40	–	–	0.83	0.92	0.10	0.02	0.02	0.32	2.21
Netherlands	0.62	0.46	0.59	–	–	0.70	0.34	0.25	0.02	0.01	0.34	1.66
Portugal	0.24	0.42	0.24	–	–	0.51	0.19	0.09	0.02	0.01	0.06	0.90
Spain	1.64	0.38	0.29	–	0.01	0.59	0.65	0.67	0.05	0.05	0.29	2.30
Sweden	0.85	0.11	0.02	–	–	0.15	0.13	0.32	0.02	0.01	0.35	0.99
United Kingdom	0.44	0.65	0.42	–	0.00	0.46	0.57	0.13	0.01	0.01	0.33	1.50
Advanced economies (post-communist EU)												
Czech Republic	1.02	5.85	0.45	–	–	1.99	4.57	0.29	0.04	0.05	0.36	7.31
Estonia	–	0.03	0.30	–	0.00	0.21	0.05	–	–	–	0.07	0.32
Latvia	1.20	0.05	0.74	–	–	0.64	0.71	0.14	0.05	0.10	0.36	2.00
Lithuania	1.94	0.14	0.77	–	–	0.84	1.47	0.21	0.10	0.09	0.13	2.84
Slovak Republic	0.20	1.37	0.65	–	–	1.17	0.76	0.07	0.02	0.01	0.19	2.22
Slovenia	–	1.37	0.14	–	–	0.60	0.94	–	–	–	(0.03)	1.51

Emerging and developing Europe (post-communist EU)												
Bulgaria	1.90	8.80	0.58	–	–	3.11	7.46	0.26	0.13	0.01	0.30	11.28
Croatia	0.38	1.12	0.65	–	–	0.70	1.12	0.04	0.02	0.01	0.26	2.15
Hungary	1.05	1.30	1.01	–	–	1.05	1.33	0.17	0.07	0.00	0.73	3.36
Poland	0.98	5.08	0.44	–	0.00	2.27	3.52	0.19	0.10	0.02	0.40	6.50
Romania	1.21	4.43	0.86	–	–	1.27	4.43	0.13	0.12	0.06	0.48	6.50
Emerging and developing Europe (post-communist non-EU)												
Albania	–	0.94	0.02	1.36	0.54	0.17	0.79	–	–	–	0.82	2.32
Bosnia and Herzegovina	–	33.53	0.11	–	–	4.16	29.47	–	–	–	0.01	33.65
FYR Macedonia	1.09	10.23	0.14	–	–	1.95	8.92	0.27	0.09	0.03	0.20	11.46
Kosovo	1.14	28.07	1.11	–	–	5.54	24.09	0.23	0.08	0.03	0.36	30.32
Montenegro, Rep. of	–	19.50	0.51	–	–	2.10	17.92	–	–	–	(0.00)	20.01
Serbia	–	32.69	0.60	–	–	4.25	28.94	–	–	–	0.09	33.29

Source: IMF country-level energy subsidies database, October 3, 2018; the full dataset can be accessed at www.imf.org/en/Topics/Environment/energy-subsidies, in the category "Data and diagnostic tools/Country-level energy subsidy estimates."

Note

Figures in brackets indicate that energy subsidies are negative. According to the IMF's definition, a negative post-tax energy subsidy arises when consumer prices are above "supply costs plus a tax to reflect environmental damage and an additional tax applied to all consumption goods to raise government revenues." (Coady et al., 2015).

Conclusion

By looking at the hidden aspects of energy poverty symptoms and drivers, our objective has been to highlight some difficulties that occur when it comes to addressing energy poverty in post-communist Europe. Our analysis of types of energy poverty symptoms – visible, measurable, and hidden – is not specific to post-communist Europe. However, the types of hidden energy poverty symptoms we identified are to some extent specific to the region. Examining these hidden symptoms allows highlighting the sometimes similarly hidden severity of energy poverty in the region.

Likewise, the distinction we propose between individual drivers and context factors is not specific to the region. But our analysis of context factors in the case of post-communist Europe reveals that these context factors might be particularly problematic when it comes to addressing energy poverty there. The five characteristics of energy poverty in the region we identified – spatial disparities in terms of energy poverty prevalence, poor targeting and segregation between social groups, historical lack of investment in housing and in infrastructures, energy independence considerations of countries, and persistence of inefficient energy subsidy systems – have been present in the countries of the region for decades. The success of energy poverty alleviation strategies will therefore depend not only on the ability of countries to propose definitions of energy poverty and to design policies dedicated to energy poor households, but also on their ability to undergo more profound reforms to address some of the structural roots of the problem.

Notes

1 www.energypoverty.eu.
2 National Energy and Climate Plan of the Czech Republic November 2019, https://ec.europa.eu/energy/sites/ener/files/documents/cs_final_necp_main_en.pdf.
3 "Energy poverty is often a consequence of poverty itself and must therefore be treated as a social policy issue." Slovak Ministry of Economy (2019) Integrated National Energy and Climate Plan for 2021 to 2030, prepared pursuant to Regulation (EU) 2018/1999 of the European Parliament and of the Council on the Governance of the Energy Union and Climate Action. Bratislava, December 2019. https://ec.europa.eu/energy/sites/ener/files/sk_final_necp_main_en.pdf.
4 An example of such a situation is discussed in the chapter on Hungary in the present book.
5 In Romania, one example was discussed in experts' meetings where households connected to the district heating system suffered from supply interruptions for two weeks in December 2018 – "100,000 apartments were cut over Christmas in Bucharest" – followed by further interruptions in January 2019.
6 The data used are drawn from the Life in Transition Survey of the European Bank for Reconstruction and Development. Their analysis focuses on 11 countries from former communist Europe, which represents 11,598 surveyed individuals (Druică et al., 2019).

7 For example Rademaekers et al. (2016) identify the following categories of context factors: (a) Structural: the socio-political system, which refers to the previous and current political system; (b) Markets, which represent the type of energy market, the extent of liberalization and level of competition; (c) Natural systems – climate, which determines energy demand; (d) Macro-economy, which can influence income; (e) Economic – income, which can influence the level of energy services provision; and (f) Policy framework, which relates to the policy framework in place to support vulnerable customers or to address energy poverty.

8 As indicated by the IMF: "Post-tax consumer subsidies exist if consumer prices for energy are below supply costs plus the efficient levels of taxation. The efficient level of taxation includes two components. First, energy should be taxed the same way as any other consumer product. Second, some energy products contribute to local pollution, traffic congestion and accidents, and global warming – efficient taxation requires that the price of energy should reflect these adverse effects on society. In most countries, taxes on energy fall far short of the efficient levels." Source: www.imf.org/en/Topics/Environment/energy-subsidies.

References

Berescu, C., Petrović, M., and Teller, N. (2012). Housing exclusion of the Roma: Living on the edge. In *Social Housing in Transition Countries* (108–123). Routledge.

Birol, F. (2007). Energy economics: A place for energy poverty in the agenda? *The Energy Journal*, 28 (3).

Boardman, B. (2010). *Fixing Fuel Poverty: Challenges and Solutions*. Earthscan.

Bollino, C. A., and Botti, F. (2017). Energy poverty in Europe: A multidimensional approach. *PSL Quarterly Review*, 70 (283).

Bouzarovski, S., and Thomson, H. (2019). *Transforming Energy Poverty Policies in the European Union: Second Annual Report of the European Union Energy Poverty Observatory*, www.energypoverty.eu/sites/default/files/downloads/observatory-documents/20-01/epov_pan-eu_report_2019_final.pdf.

Bouzarovski, S., and Tirado Herrero, S. (2017). Geographies of injustice: the socio-spatial determinants of energy poverty in Poland, the Czech Republic and Hungary. *Post-Communist Economies*, 29 (1), 27–50.

Bouzarovski, S., and Petrova, S. (2015). A global perspective on domestic energy deprivation: Overcoming the energy poverty–fuel poverty binary. *Energy Research and Social Science*, 10, 31–40.

Bouzarovski, S. and Thomson, H. (2019). *Transforming Energy Poverty Policies in the European Union: Second Annual Report of the European Union Energy Poverty Observatory*. www.energypoverty.eu/observatory-documents/transforming-energy-poverty-policies-european-union-second-annual-report.

Bouzarovski, S., Petrova, S., and Sarlamanov, R. (2012). Energy poverty policies in the EU: A critical perspective. *Energy Policy*, 49, 76–82.

Buzar, S. (2007). *Energy Poverty in Eastern Europe: Hidden Geographies of Deprivation*. Ashgate Publishing, Ltd.

Coady, D., Parry, I., Sears, L., and Shang, B. (2015). How large are global energy subsidies? WP/15/105, www.imf.org/en/News/Articles/2015/09/28/04/53/sonew070215a.

Druică, E., Goschin, Z., and Ianole-Călin, R. (2019). Energy poverty and life satisfaction: Structural mechanisms and their implications. *Energies*, 12 (20), 3988.

Dubois, U., and Meier, H. (2016). Energy affordability and energy inequality in Europe: Implications for policymaking. *Energy Research and Social Science*, 18, 21–35.

Dubois, U., Assimakopoulos, M., Biermann, P., Gouveia, J., Karlessi, T., Kyprianou, T., Mattioli, J., Murauskaite, L., and Sinea, A. (2018). Towards harmonised regional energy poverty assessments, in: Bouzarovski, S. (Ed.) (2018) *European Energy Poverty. Agenda Co-Creation and Knowledge Innovation (2017–2021)*, Policy Brief no. 1, December 2018.

European Union (2019). Special Eurobarometer 492 report, *Europeans' Attitudes on EU Energy Policy*.

Gangale, F., and Mengolini, A. (2019). *Energy Poverty Through the Lens of EU Research and Innovation Projects*. EUR 29785, Publications Office of the European Union, Luxembourg.

Graham, C. (2006). The politics of necessity: Electricity and water in Great Britain. *Journal of Consumer Policy*, 29 (4), 435–448.

Hegedüs, J. (2011). Social safety nets, the informal economy and housing poverty: Experiences in transitional countries. *European Journal of Homelessness*, 5 (1), 15–26.

Heidegger, P., and Wiese, K. (2020). *Pushed to the Wastelands: Environmental Racism Against Roma Communities in Central and Eastern Europe*. Brussels: European Environmental Bureau.

Hills, J. (2012). *Getting the Measure of Fuel Poverty: Final Report of the Fuel Poverty Review*.

Kakalejclkova, Z. (2017). *How to Improve Residential Energy Efficiency in South Eastern Europe and CIS: Policy Discussion Brief for National Governments and International Organizations*. Habitat for Humanity International. Europe, Middle East and Africa. April 2017.

Kósa, K., Daragó, L., and Ádány, R. (2011). Environmental survey of segregated habitats of Roma in Hungary: A way to be empowering and reliable in minority research. *The European Journal of Public Health*, 21 (4), 463–468.

Lenz, N. V., and Grgurev, I. (2017). Assessment of energy poverty in new European Union member states: The case of Bulgaria, Croatia and Romania. *International Journal of Energy Economics and Policy*, 7 (2), 1–8.

Lis, M., Miazga, A., and Sałach, K. (2016). *Location, Location, Location. What Accounts for Regional Variation of Fuel Poverty in Poland?* No. 09/2016. Instytut Badan Strukturalnych.

Murafa, C., Sinea A., Jiglău, G., and Bădescu, G. (2017). *Energy Poverty and the Vulnerable Consumer. How Far Are We from Europe?* Theoretical, legislative framework and statistical evidence. The Center for the Study of Democracy, Cluj.

Peneva, T. (2014). *Energy Poverty: The Bulgarian Case*. Washington, DC: International Association for Energy Economics.

Pittini, A., Koessl, G., Dijol, J., Lakatos, E., Ghekiere, L., and Goudis, M. (2017). *The State of Housing in the EU*. Housing in Europe review.

Primc, K., Slabe Erker, R., and Majcen, B. (2019a). Constructing energy poverty profiles for an effective energy policy. *Energy Policy*, 128, 727–734.

Primc, K., Slabe Erker, R., and Majcen, B. (2019b). Energy poverty: A macrolevel perspective. *Sustainable Development*, 27 (5), 982–989.

Pye, S., Dobbins, A., Baffert, C., Brajković, J., Grgurev, I., Miglio, R. D., and Deane, P. (2015). *Energy Poverty and Vulnerable Consumers in the Energy Sector Across the EU: Analysis of Policies and Measures*. INSIGHT_E.

Rademaekers, K., Yearwood, J., Ferreira, A., Pye, S., Hamilton, I., Agnolucci, P., Grover, D., Karásek, J., and Anisimova, N. (2016). *Selecting Indicators to Measure Energy Poverty*. Rotterdam, The Netherlands: Trinomics.

Robic, S., Živčič, L., Kisyov, P., Tkalec, T., and Ilievski, Z. (2016). *Energy Poverty in South-East Europe: Challenges and Possible Solutions*. www.enu.hr/wp-content/uploads/2017/02/Policy-recommendations_SEE_EU.pdf.

Rutkowski, J., Sałach, K., Szpor, A., and Ziółkowska, K. (2018). *How to Reduce Energy Poverty in Poland?* No. 01/2018. Instytut Badan Strukturalnych.

Sareen, S., Thomson, H., Herrero, S. T., Gouveia, J. P., Lippert, I., and Lis, A. (2020). European energy poverty metrics: Scales, prospects and limits. *Global Transitions*, 2, 26–36.

Soaita, A. M. (2014). Overcrowding and "underoccupancy" in Romania: A case study of housing inequality. *Environment and Planning A*, 46 (1), 203–221.

Stępniak, A., and Tomaszewska, A. (2013). *Ubóstwo energetyczne a efektywność energetyczna*. Analiza problemu i rekomendacje, Instytut na rzecz Ekorozwoju, Warszawa.

Teschner, N., Sinea, A., Vornicu, A., Abu-Hamed, T., and Negev, M. (2020). Extreme energy poverty in the urban peripheries of Romania and Israel: Policy, planning and infrastructure. *Energy Research and Social Science*, 66, 101502.

Thomson, H., and Snell, C. (2013). Quantifying the prevalence of fuel poverty across the European Union. *Energy Policy*, 52, 563–572.

Thomson, H., Bouzarovski, S., and Snell, C. (2017). Rethinking the measurement of energy poverty in Europe: A critical analysis of indicators and data. *Indoor and Built Environment*, 26 (7), 879–901.

Velte, D., Magro, E., and Jiménez, I. (2010). ReRisk. *Regions at Risk of Energy Poverty*. Applied Research Project 2013/1/5. Luxembourg: ESPON.

Part I

Energy poverty and politics

2 Trapped in politics

Energy poverty in Hungary

Anna Zsófia Bajomi, Nóra Feldmár, and Lea Kőszeghy

Introduction

In Hungary, the cost of energy has been a highly politicized topic for decades. In political discourse, the focus has been on the general affordability of energy for Hungarian households, without much consideration of more detailed social issues. For decades energy prices have been state regulated and generally framed as an instrument to "protect" households from being exposed to utility providers. The strong focus of political discourse on energy affordability perhaps distinguishes Hungary from other countries in the region. At the same time there is a high level of energy poverty in the country resulting from interplay among social, economic, and political factors, and this energy poverty is imperfectly addressed.

In this chapter we (1) provide an overview of existing information on the prevalence and characteristics of energy poverty in Hungary; (2) offer an understanding of the larger structural and contextual causes shaping energy poverty with due regard to the specific post-socialist and post-transition traits; as well as (3) analyze public policies, showing how welfare and energy policies have not only failed to effectively address energy poverty so far but have also generated harmful outcomes.

Energy poverty in Hungary: the current state of knowledge

Existing literature on energy poverty in Hungary

Energy poverty in Hungary has not been at the forefront of academic research and policy-making until recently, which makes assessment of the phenomenon difficult. There have been relatively few attempts to characterize energy poverty in the Hungarian context. This section reviews the existing academic literature on the topic.

The first comprehensive assessment of energy poverty in Hungary was published in 2010, by Tirado Herrero and Ürge-Vorsatz, who collected and analyzed various indicators related to energy poverty, such as energy costs to income ratios and excess winter deaths (Herrero and Ürge-Vorsatz, 2010, p. 28).

Their work also highlighted that the energy use in Hungarian households is among the highest in the European Union, ranked tenth out of 27 at the time. In addition to a quantitative exploration of energy poverty, the authors presented case studies on various forms of energy poverty in Hungary along with an overview of the groups that are primarily affected, such as Roma households living in remote rural areas, district heating consumers (as a specific post-communist legacy), and elderly households. Regarding energy supply, Herrero and Ürge-Vorsatz highlighted the difficult situation of energy providers caused by high rates of non-payment of bills. Finally, the authors recommended that the government switch away from "supply-side approaches based on consumption of energy to demand-side solutions focused on the provision of energy services (e.g., ensuring thermal comfort, covering lighting requirements, etc.)" to tackle energy poverty in the long term (Herrero and Ürge-Vorsatz, 2010, p. 31).

In their analysis of energy vulnerability and its "wider institutional and spatial configurations," Bouzarovski et al. (2014) take Hungary as an example to "illustrate the systemic driving forces and implications of domestic energy deprivation" (Bouzarovski et al., 2014, p. 1). In their analysis, the authors provide a systematic overview of structural factors in Hungary, including "systemic path-dependencies" and structural factors inherited from the communist centralized state. Single-family houses built before 1990 were examined as an illustration of the consequences of the structural heritage of the past. These structures represent the majority of the dwelling stock and are extremely energy inefficient. These vulnerability factors result in various household coping strategies such as delays in payments and fuel switching. Their study also offers an overview of government policies related to price regulation (Bouzarovski et al., 2014).

Further literature highlights additional elements that contribute to energy poverty in Hungary, such as the high proportion of low-income groups occupying the less-efficient dwelling types (Ministry of National Development, 2015a), heating with solid fuels in inefficient stoves, and the possibility of disconnection from energy services (Bajomi and Feldmár, 2018). All of these situations might result in under-consumption. The lower energy spending of affected households is not reflected in energy poverty indicators such as the double median (2M) or the low-income high-cost (LIHC) indicator. Other streams of literature related to housing research have also analyzed questions related to energy poverty in Hungary. Here, the focus is mainly on affordability, housing, and energy subsidies (Hegedüs et al., 2009; Kováts, 2015a; Misetics, 2018, 2013). This literature highlights the concentration of affordability problems in detached (family) housing and in the cases of single-parent households and households with children and other family members (Hegedüs and Somogyi 2018). It also shows the ineffectiveness of the existing subsidy system in terms of its ability to respond to affordability problems in a targeted and effective manner.

Beyond the academic literature, several energy poverty analyses have been published by Habitat for Humanity Hungary. The organization has been mapping available data concerning energy poverty in its annual report on housing poverty in Hungary each year since 2011 (Bajomi and Feldmár, 2018; Balogi et al., 2014; Koltai, 2015; Kőszeghy, 2017). In a work on deaths related to hypothermia, Koltai emphasizes the concentration of hypothermia deaths in energy poor single households (Koltai, 2013). An analysis of the recent solid fuel subsidy (Bajomi, 2018) highlights problems concerning the adequacy of the subsidy and its targeting: on the one hand, beneficiaries tend to receive a higher amount of fuel in higher income localities, while on the other hand households that live in larger settlements are not eligible, regardless of their needs.

Energy poverty indicators

Data concerning the scale and types of energy poverty in Hungary are fragmented and often difficult to obtain. This section summarizes available data based on two approaches: households' experiences and the proportion of energy expenditures in household incomes.

For experience-based indicators, 2018 Eurostat data show that 23 percent of the population of Hungary lived in a dwelling with a leaking roof, damp walls, floors or foundations, or rot in the window frames of floor (Eurostat, 2018c). Dwellings of the poorest condition, quality, and efficiency are typically inhabited by low-income families and individuals without any savings to invest in improvements (Ministry of National Development, 2015b). According to Eurostat, in 2018, 6 percent of the population was unable to keep their homes adequately warm (Eurostat, 2018a), and 11 percent of households have been unable to pay utility bills on time in the past 12 months (Eurostat, 2018b). Households below the poverty line are significantly more affected: 44 percent of people living in poor households resided in housing with the aforementioned housing quality problems, 20 percent were unable to keep their home adequately warm, and 29 percent were in arrears on utility bills (Eurostat, 2018a, 2018b, 2018c). In 2018 utility arrears were nearly twice as prevalent among those who live in thinly populated areas (14.4 percent) than among those living in densely populated territories (7.9 percent) (EU Energy Poverty Observatory, 2018).

National data exist on households being disconnected from energy supplies. According to Hungarian regulations, consumers can be cut off from gas, energy, and district heating provision when they have been in arrears for over 60 days. In 2018 approximately 3 percent of gas users (96,000 households) and 10 percent of electricity users had over 60 days of arrears. In 2016, 45,000 households were cut off from gas or electricity services (MEKH, 2019).

A further indicator of energy poverty is excess winter mortality. According to an initial, rough estimate made by Herrero and Ürge-Vorsatz (2010),

"In Hungary, between 1400 and 2400 people (between 25% and 44% of all [excess winter deaths]) may die prematurely every year because of poor housing and living conditions related to fuel poverty" (Herrero and Ürge-Vorsatz, 2010, p. 28). According to the Central Statistical Office, between 1998 and 2018 on average 238 persons died every year – amounting to 4,772 persons in 20 years – because of exposure to excessive cold in their homes (Central Statistical Office, 2019a). This statistic measures mainly mortality caused by hypothermia, which is only one cause of excess winter mortality; other causes not included in that statistic are respiratory and circulatory conditions.

Concerning the proportion of energy expenditures in households' net incomes, Figure 2.1 shows that in 2017, the average share of energy costs was 11.5 percent. This proportion is significantly higher in lower-income groups: households in the lowest income quintile spend 14.8 percent of their net income on energy, while households in the richest quintile spend 7.4 percent on energy (own elaboration based on Central Statistical Office, 2018a and 2018b, and Central Statistical Office, 20181 and 2018b). Households in the lowest income decile have virtually no remaining income after essential expenditures. Due to the level of economic strain, the risk of accumulating arrears is very high in this group.

More disaggregated statistics show that the income share of energy costs follows distinct spatial patterns: households living in villages have higher nominal energy expenditures and a higher share of energy spending than

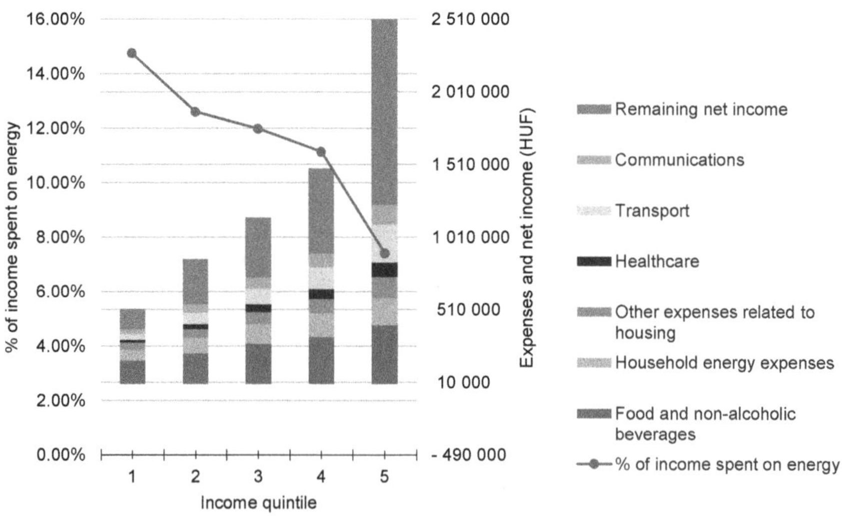

Figure 2.1 Annual expenditure on essential goods and services and remaining net income by income quintiles 2017.

Source: Compiled by the authors based on data from the Central Statistical Office (2018a and 2018b).

households that live in towns or larger cities. Households living in the capital had to dedicate the lowest share of their income to pay energy bills at 9 percent, while households living in villages spent 14 percent (Central Statistical Office, 2018a).

Fülöp and Lehoczki-Krsjak (2014) provide estimates of energy poverty based on the 2012 Household Budget Survey. They estimated that 10 percent of Hungarian households were energy poor according to the 2M indicator. The energy poverty rate was 21 percent with a measurement based on the LIHC indicator. The socio-economic characteristics of households identified by the 2M and LIHC indicators are different: while the 2M indicator preferentially captures single-person households, the LIHC indicator captures households with children as energy poor in above-average proportions.

More recently Kőszeghy and Feldmár (2019) measured energy poverty based on the 2015 housing survey of the Central Statistical Office. They estimate the proportion of energy poor households to be 10.5 percent based on the 2M indicator, and 14.4 percent on the LIHC indicator (404,000 and 556,000 households, respectively). For the first time, their calculations tried to grasp under-consuming households with the M/2 indicator (low absolute energy expenditure indicator). According to that indicator, 10.4 percent of Hungarian households (403,000 households) are under-consuming energy.

Different indicators for energy poverty capture different socio-economic groups. With the 2M indicator, one-person and single-parent households in the first two income quintiles, especially with low educational attainment levels, and a household head over 50 years of age, are present in a significantly above-average proportions. The LIHC indicator, instead, captures households with children in the first three income quintiles, also with low educational attainment levels, but a somewhat younger household head, between 30 and 50 years. Both the 2M and LIHC indicators capture rural households in smaller localities and in detached housing in above-average proportions. The "hidden" energy poor households identified by the M/2 indicator significantly differ from both: according to this indicator measuring under-consumption, single-person households in the first income quintile are classified as energy poor in significantly above-average proportion. Moreover, data suggest that under-consuming households are a heterogenous group, present both in rural areas in small localities and in the capital, Budapest, and its surroundings, with old (above 70 years of age) and young (below 30 years of age) household heads. Tenants in urban areas appear as one of the under-consuming groups as well (Kőszeghy and Feldmár, 2019). Differences between households identified by different energy poverty indicators reflect the multifaceted nature of energy poverty and the need to develop tailored policies responding to the needs of different groups of energy poor households. This is especially true of spatial differences: the predominance of rural households in energy poor households according to the 2M and LIHC indicators, and the appearance of under-consuming urban households identified by the M/2 indicator.

Contextual and structural factors shaping energy poverty in Hungary

As defined by Bouzarovski et al. (2014, p. 276), energy poverty can be seen as "a situation in which a household lacks a socially and materially necessitated level of energy services in the home." A main cause of energy poverty in Europe is a combination of "low income with thermally inefficient homes." Further relevant factors determining energy poverty include "household size, gender, class or education" (Bouzarovski et al., 2014, p. 279). Certain analyses also identify the role of ethnicity (Reames, 2016).

This section reviews the contextual and structural factors shaping energy poverty in Hungary. We analyze the roles of socio-economic inequality, housing poverty, and energy services and prices. We also examine the specific situation of the Roma, a group suffering from particularly high levels of energy poverty.

Socio-economic inequalities

Since the regime change in 1989, the gap between rich and poor has been widening in Hungary. Though in international comparison, social inequalities in Hungary are relatively mild, the highest and lowest income groups are detached from other parts of the population, and social mobility is weak (Tóth and Szelényi, 2018). A significant proportion of the Hungarian population struggles to meet basic needs – including energy needs – while the poorest face severe deprivation.

Following the regime change, the centrally planned economy of socialist Hungary transitioned to a market economy and its politics to a multi-party democratic system. The winners of the transition mainly came from the leading elites of the previous system and the losers mainly from skilled workers and laborers (Ferge, 1996). Unemployment went from virtually non-existent to a peak of 12.1% in 1993, leaving 500,000 people without jobs at that time (Central Statistical Office, 2010).

From 1989 to the present, policies and market mechanisms have not been able to improve the situation of the millions who found themselves in a vulnerable position. After the 2008 crisis, austerity measures, tax reforms, and gradual weakening of social services created a widening gap between the rich and the poor. As Figure 2.2 shows, the disposable income of the first decile (the poorest 10 percent) increased by 12 percent between 2010 and 2017, while the disposable income of the last decile (the richest 10 percent) increased by 28 percent (adjusted for inflation).

Income inequality also shapes the face of energy poverty. As Figure 2.2 shows, the net income of the lowest income decile of households remaining

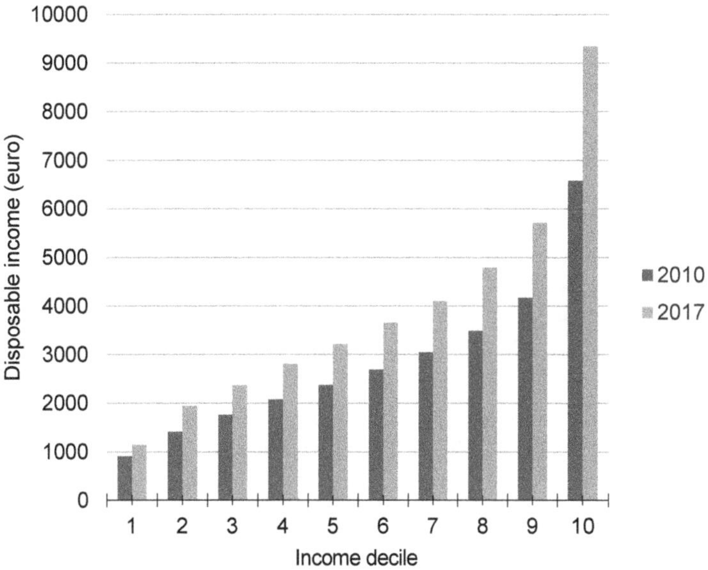

Figure 2.2 Yearly disposable income per capita in euro by income deciles in 2010 and 2017.

Source: Compiled by the authors based on data from the Central Statistical Office (2018a).

after basic expenditures is a factor of 12 lower than that of the highest income group. Households of the lowest income quintile spend an average of 17 percent of their income on energy, while energy costs weigh less than half that much in the income of the highest income decile (8 percent).

A useful entry point to understand inequality is to consider indicators of material poverty. In 2018 single adults with dependent children had the highest risk of poverty (31.5 percent), whereas the risk of poverty in the larger group of all households with children was 10.6 percent in 2018, a substantial drop from the 16.3 percent recorded in 2009. The share of households without children facing poverty nearly doubled from 2009 to reach 12.5 percent in 2018 (Eurostat, 2018d). According to another source, one in three (33.6 percent) children are at risk of poverty or social exclusion. On this last criterion, Hungary is ranked the fourth-lowest country in the EU (European Union Agency for Fundamental Rights, 2018). As we will discuss in a later section, the dynamics of inequality and poverty are strongly affected by social and fiscal policies.

Housing poverty

The current state of the housing stock and prospects for refurbishment are strongly determined by the "super-homeownership" nature of the Hungarian housing system: approximately 90 percent of the population lives in owner-occupied housing. Public housing constitutes less than 3 percent of the housing stock, while 6 to 8 percent of dwellings are private tenements. Owner occupation is the dominant tenure form in all income groups. Even in the lowest income decile 77 percent of households live in owner-occupied housing (Central Statistical Office, 2018b). As a result, many energy poor people are homeowners with low incomes and a lack of savings, leaving them unable to maintain and improve their homes.

In Hungary 20 to 30 percent of the population faces housing deprivation, including problems related to the affordability, quality, and/or energy efficiency of the home, insecurity of tenure, or location problems (Pósfai, 2018). According to Eurostat, 7.5 percent of Hungary's total population and 27.3 percent of its children are in severe housing deprivation (Eurostat, 2018d).

For tenants, access to social housing is very limited, as less than 3 percent of the total housing stock is public (Central Statistical Office, 2018c) and only about half of it – sixty-seven thousand dwellings – is rented on a social basis (Central Statistical Office, 2016a; Eurostat, 2018c). Tenants renting either public or private dwellings face various affordability problems: according to the 2015 Budget Housing Survey of the Central Statistical Office, 62 percent of public tenants and 53 percent of private tenants have some kind of affordability problem (low residual income after housing costs, substandard housing due to inability to afford a higher quality dwelling, or housing costs over 40 percent of household income).

For homeowners, the 2008 financial crisis put major pressure on households with mortgages, on the one hand because of the dramatic change in the foreign-exchange rates, and on the other hand many households lost their jobs and saw their incomes decrease during this crisis. At that time, 22 percent of households had a mortgage. They had to deal with the repayment of drastically increasing installments. As a result, nearly 10 percent of Hungarian households were faced with major difficulties in paying back their credit. In 2015, 45 percent of owners with a mortgage were still facing some form of affordability problem (Hegedüs and Somogyi, 2018).

Housing affordability problems have also increased since the early 2010s due to booming property prices and an increase of private rental fees. At the same time, housing and equipment quality problems – the other key factor of energy poverty – still prevail in the Hungarian housing stock. It has been estimated that 80 percent of dwellings do not meet current energy efficiency standards (Ministry of National Development, 2015b). This situation is to a large extent due to historic reasons. Hungary has a high proportion of pre–World War II housing, as well as self-constructed detached houses and housing constructed with industrial technologies between the 60s and 80s. These dwellings lack effective heat insulation and there have been no comprehensive refurbishment programs addressing these problems in the past decades.

The Roma population

Energy poverty is particularly prevalent in the Roma population, which is disproportionately affected by housing poverty and poverty in general. Roma people represent between 3 and 8.5 percent of the Hungarian population (Central Statistical Office, 2016b; Pénzes et al., 2018). A majority of Roma face severe difficulties, such as multiple forms of discrimination, educational segregation, low positions on the job market, and poor health and housing conditions (Bernát, 2014; European Commission, 2015). In 2016, 55.5 percent of Roma citizens lived in severe material deprivation (Central Statistical Office, 2016c) and their housing situation is historically much worse than that of the majority of the Hungarian population (Kemény and Janky, 2004; Lengyel, 2006; Teller, 2011). According to 2011 data (Teller, 2011), 21 percent of Roma people do not have access to piped water and 32 percent have no access to flush toilets, while the corresponding proportions for the total population were 2 percent and 6 percent, respectively (Hungarian Central Statistical Office, 2011). In the early 2010s, 32.1 percent of Roma people lived in a home with leaking roof, 49.5 percent in one with wet walls, 40.7 percent had rotten windows or door frames, and 27.4 percent had dwellings that did not get enough light (Letenyi and Varga, 2011). According to 2011 United Nations Development Programme data, 81 percent of Roma people in Hungary live in a household using firewood as a heating source (FRA and UNDP, 2012).

Geographical segregation is a major issue in Hungary: according to a 2010 data collection, approximately three hundred thousand people (3 percent of the total population) lived in low-quality dwellings located in ethnically segregated areas (Ministry of Human Resources, 2015). The Roma people are particularly affected: according to a 2010 survey, 21.8 percent of Roma households lived in segregated slums outside of villages and towns. Furthermore, one hundred to two hundred villages in remote disadvantaged regions became "ghetto-villages" (Váradi and Virág, 2014). These segregated areas tend to have poorer access to amenities such as concrete roads, gas pipes, running water, and sewage systems. Poor housing quality is an important energy poverty factor here, as dwellings in these areas are more likely heated with low-performing individual heaters fueled by wood, coal, or other materials.

Energy services and prices

Having explored the larger socio-economic perspective, we now turn to energy services and energy prices. To assess the specificities of Hungary in those respects, it is necessary to understand the main characteristics of the Hungarian energy sector.

The energy sector in Hungary

Hungary does not have significant reserves of fossil fuels, and therefore a majority of the primary energy use is derived from imported energy carriers; for example, 57 percent in 2017 (MEKH, 2019). In addition to fossil fuels, nuclear power (produced in Hungary's nuclear power station in Paks) and biomass combustion account for a significant share of domestic consumption. Figure 2.3 shows the development of Hungary's energy mix between 1990 and 2018.

Domestic electric power generation is primarily characterized by centralized fossil and nuclear power plants: 50 percent of the electricity is generated by the Paks Nuclear Power Plant and 17 percent by the lignite-fueled Mátra Power Plant, while almost 30 percent of total consumption comes from imports (MEKH, 2019). Primarily as a result of EU regulations to reduce carbon emissions, many power plants have been partially or completely converted to biomass combustion and in recent years a few solar power plants have been built.

Production of thermal energy is predominantly (68 percent) from gas, of which only 17 percent is supplied from domestic sources (MEKH, 2019), the rest being imported, mainly from Russia. Household heating is primarily accomplished by burning gas supplied through gas pipelines, and to a lesser extent by using district heat (steam) produced in gas, coal, or biomass power plants. Firewood also plays a very important role in the heating of households.

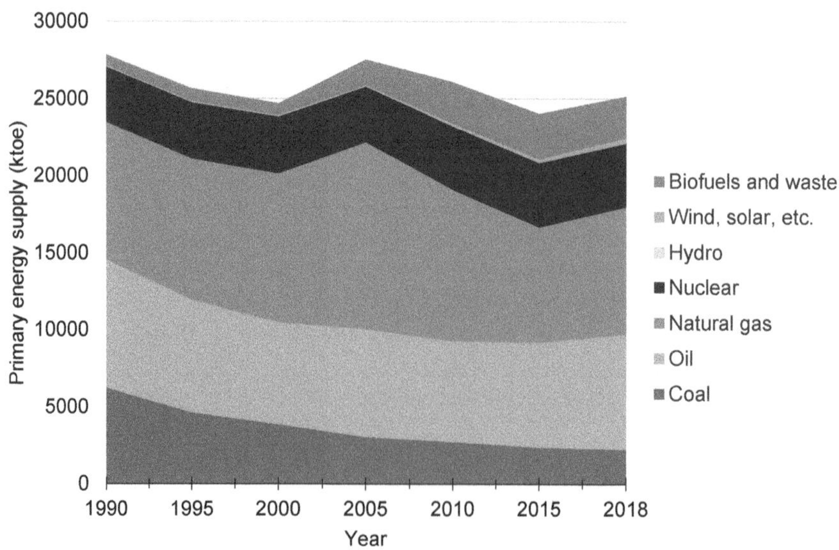

Figure 2.3 Total primary energy supply by source, Hungary 1990–2018, ktoe.
Source: IEA (2019).

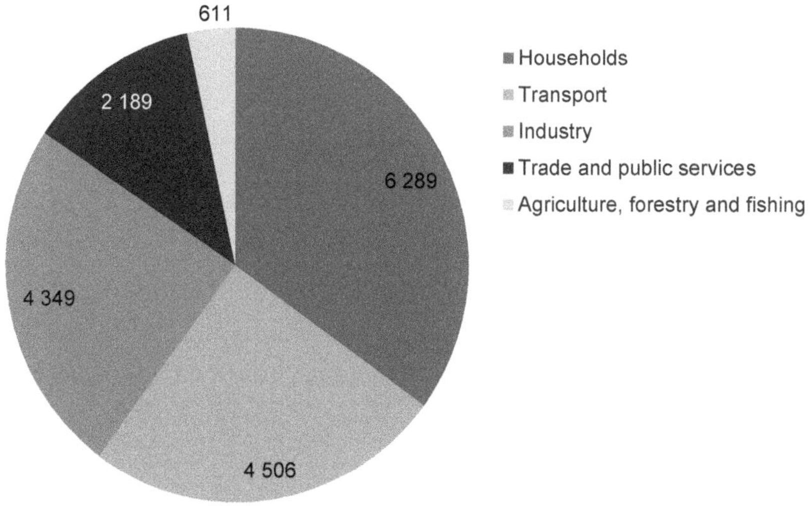

Figure 2.4 Energy consumption by end-use sectors (terajoule) 2017.
Source: Compiled by the authors based on data from the MEKH (2019).

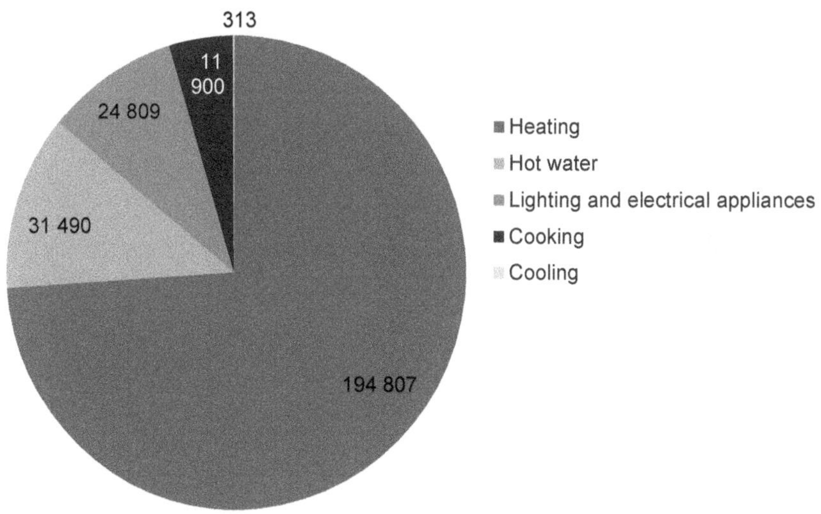

Figure 2.5 Energy consumption in households by type of end use (terajoule) 2017.
Source: MEKH (2019).

Although Hungary's renewable energy potentials are relatively good, only solid biomass sources (mainly firewood) are currently being widely exploited; 81 percent of the renewable energy used in Hungary comes from solid biomass. In 2017, a total of 12 percent of primary energy use was supplied from renewable sources, while only 1 percent was derived from non-combustible renewable sources such as wind, solar, hydro, and geothermal (MEKH, 2019).

Households account for 35 percent of the energy consumption in Hungary (Figure 2.4). This includes heating and cooling of residential buildings, domestic hot water supply, energy used for cooking and lighting, and consumption by other household appliances and equipment. Space heating represents 75 percent of households' energy consumption (Figure 2.5).

Uneven access to energy services

As gas is the most widely used fuel in Hungary, it is important to understand disparities in the accessibility of the gas network. The gas network in Hungary reaches 91.2 percent of localities with 72.9 percent of households consuming piped natural gas (Central Statistical Office, 2017). Some households live in localities with no or only partial access to gas; these are usually rural areas. They have practically no other choice than to heat with solid fuels, since the only alternative would be heating with electricity, which is highly expensive. Installing cleaner energy sources, such as solar panels, is not easily affordable for most households.

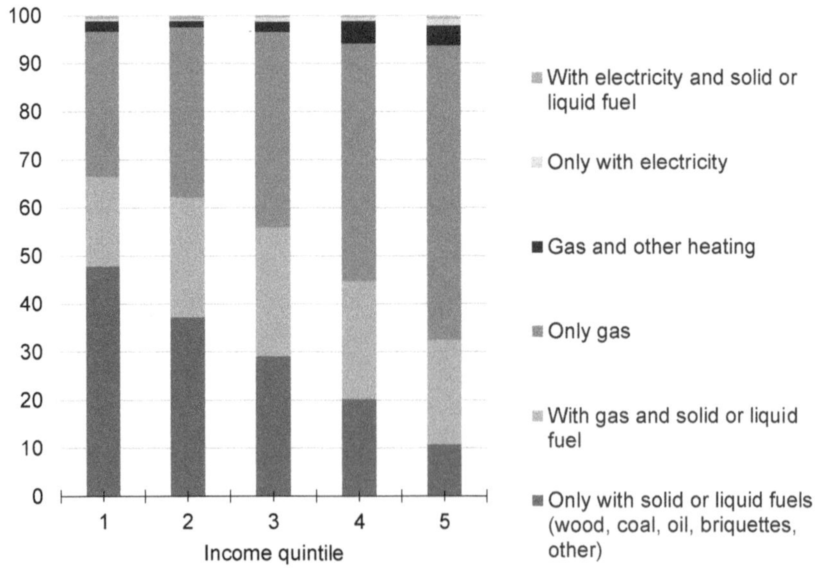

Figure 2.6 Share of households by source of heating by income quintiles 2016.

Source: Compiled by the authors based on data from the Central Statistical Office (2016c, p. 46).

As Figure 2.6 shows, the lowest income group is heavily dependent on the use of solid fuels. The higher the income, the higher the households' reliance on gas. Synchronous use of solid fuels and gas is also common (Bouzarovski et al., 2016). A prior study highlighted that a growing number of households are switching from gas to solid fuels as part of households' strategy to mitigate difficulties deriving from rising energy costs. The reason of fuel switching could be predominantly fuel price sensitivity, but there are also high costs for installing modern gas heaters as well as situations where gas services were disconnected due to debts (Bucsky, 2018). Extremely poor households that have been disconnected often lack resources and capacities to go through the expensive and complex process of debt repayment and reconnection. In addition to uneven access to physical services, there are important inequalities in terms of access to social services that could prevent and handle indebtedness. In smaller, remote villages the availability of such services is much poorer.

Energy prices and subsidies

In the state-socialist era energy prices were heavily subsidized as a result of governmental efforts to maintain a certain standard of living for households. After the regime change in 1989, the energy industry underwent major restructuring and privatization and the heavy state subsidization of energy prices ceased.

Since 1990, household energy prices have skyrocketed (Figure 2.7) representing the largest increase in prices among all basic goods and services

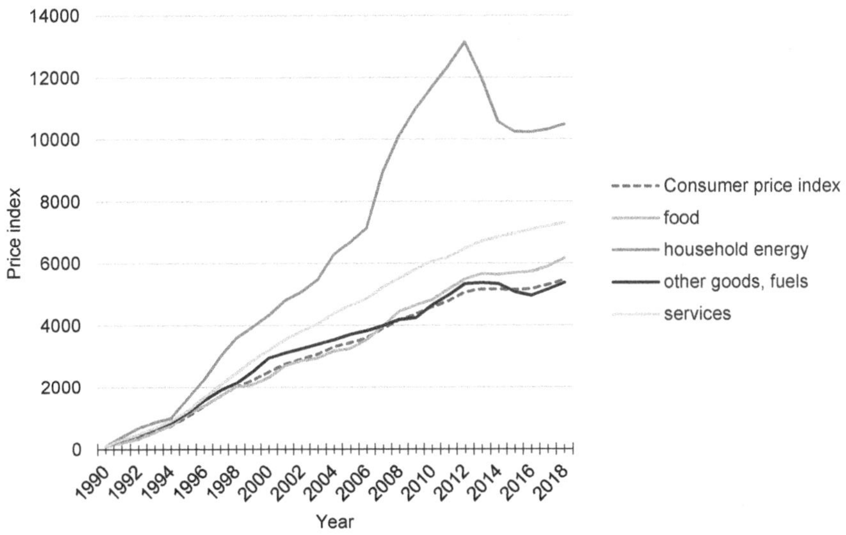

Figure 2.7 Consumer Price Indexes.

Source: Compiled by the authors based on data from the Central Statistical Office (2019b).

(Kasnyik, 2015). This increase in energy prices resulted in serious housing affordability problems for households (Hegedüs et al., 2009). In the 2000s, one fifth of households spent more than one-third of their incomes on housing. After the crisis of 2008, the housing-cost overburden rate – defined as the share of households that spend more than 40 percent of their income on housing – increased and reached its peak in 2013 at 14.3 percent (Eurostat, 2020).

In 2012, this trend toward very high housing-cost burden was halted by a stark governmental intervention. Measures of the so-called "Utility Price Reduction Act" were implemented, forcing utility companies to reduce their prices by 10–25 percent overall compared to prices in 2012 (Act 2013/LIV on the Enforcement of Utility Cost Reduction, 2013). This intervention reduced the price of energy for all domestic consumers by the same percentage. This resulted in households with higher levels of energy consumption achieving higher absolute savings on their bills (see Figure 2.8). Moreover, the scheme did not cover energy sources used by many low-income households, most notably firewood. The technical reason behind this is that solid fuels are not distributed through energy companies. The regulatory measures that stipulated the price cut, executed relatively "easily" through the energy regulator, might have been more difficult to apply in the sector of wood and coal, where small retailers operate. Also, setting wood and coal prices at a centrally set level would be legally questionable.

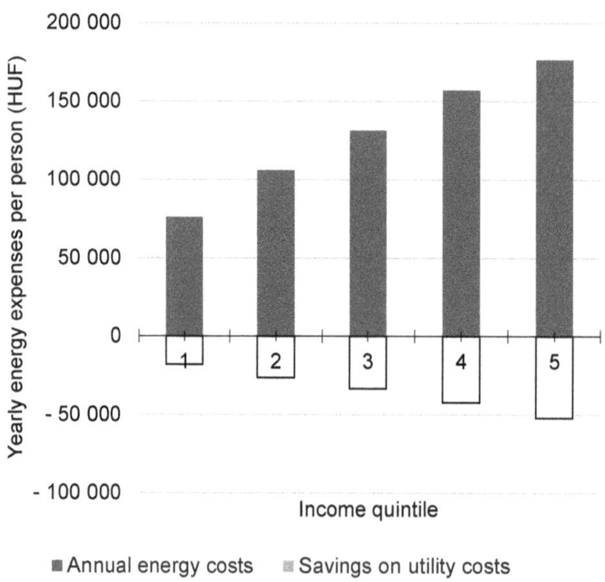

Figure 2.8 Yearly energy expenses per person and savings after utility cost reduction compared to 2012 prices, by income quintile.

Source: Compiled by the authors based on data from the Central Statistical Office (2018b).

Currently, this utility price cut is being examined by the European Court of Justice, as the European Commission found it contrary to EU law (European Commission, 2018).

More recently, in 2018, the government offered a 12,000 HUF (37 euros at 2018 currency exchange rates) "winter" subsidy for district heating (Government of Hungary, 2018b) and gas consumers (Government of Hungary, 2018b). It also provides compensation of 12,000 HUF for households who have no access to gas or district heating (Government of Hungary, 2018b). However, while the winter subsidy was automatically integrated into the utility bills of gas and district heating consumers, users of wood had to apply for the subsidy at the local government level.

Compared to other EU member states, Hungary currently has one of the lowest household energy consumer prices. Electricity is cheaper only in Lithuania and Bulgaria and gas is cheaper only in Romania (Eurostat, 2018e). However, when compared in terms of purchasing power parity, it can be seen that the Utility Price Reduction Act brought consumer energy prices down to the EU average from the previously much higher level (see Figures 2.9a and 2.9b).

Solid fuel heating

Solid fuel heating is extremely widespread in Central and Eastern European countries and is tightly linked to energy poverty in the region. In Hungary, 42.3 percent of households use solid fuels to some extent for heating or cooking; that figure comprises 1.74 million households and over 4 million individuals (Central Statistical Office, 2016c). Solid fuels are used to heat 36 percent of dwellings at least partially and 21 percent exclusively (Hungarian Central Statistical Office, 2011). As shown in Figure 2.6, solid fuels (including wood, briquettes, pellets, and coal) are the predominant heating source of lower-income households.

Low-income households who rely only on solid fuels often lack resources to invest in cleaner and more-efficient heating technologies. Therefore, the predominant mode of heating for low-income families is by individual solid fuel space heaters, i.e. woodstoves. Most of these are inefficient and have a high rate of polluting emissions (especially PM10 and PM2.5). Combined with the lack of insulation in the houses and the poor quality of the fuel – freshly cut "wet" wood, lignite, and often synthetic waste materials – this leads to a low efficiency of heat provision in the home.

In addition, the price of wood has been continually rising (Figure 2.10) due to various factors, such as the decrease of imports from Romania (Dobrescu, 2017), the Ukrainian prohibition on wood exports (European Union Delegation to Ukraine, 2019), and the increase in labor costs in Hungary. The latter resulted in an increase in the prices of labor-intensive goods, including wood (Government of Hungary, 2019, p. 30). Consequently, many low-income households relying on wood as their energy

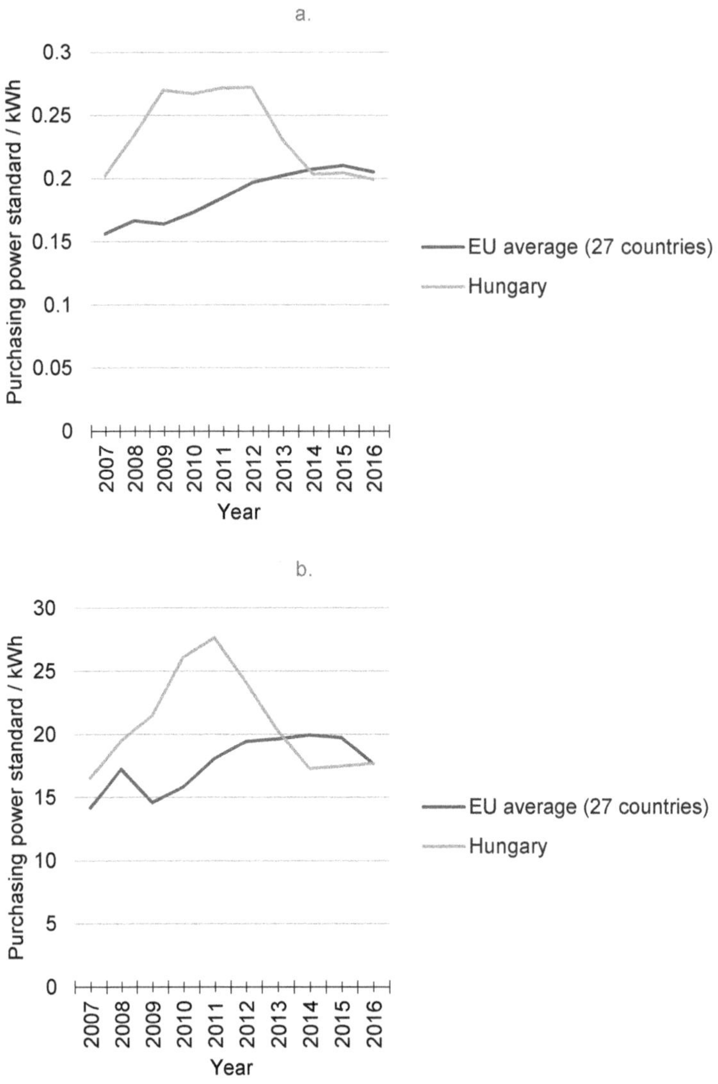

Figure 2.9 Electricity (a) and gas (b) prices for household consumers in the EU in terms of purchasing power standard (all taxes and levies included), Consumer Price Indexes.

Source: Compiled by the authors based on data from Eurostat (2019).

source went into debt and sometimes resorted to illegally sourcing wood and burning synthetic materials such as plastic waste, discarded furniture, or clothes. The air pollution effects and the health damaging effects of these practices are extremely high.

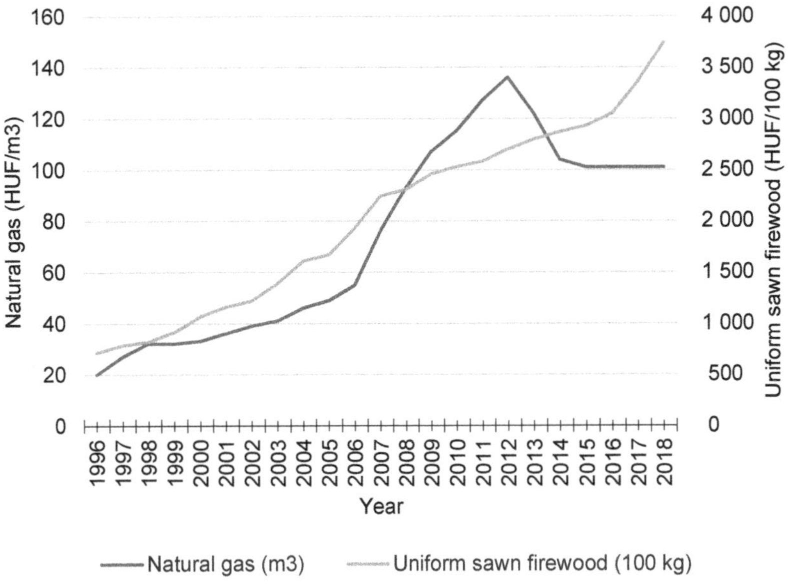

Figure 2.10 Price of natural gas and firewood 1996–2018.

Source: Own elaboration, data retrieved from Central Statistical Office (2019c).

Policy and politics

The contextual and structural factors discussed above significantly affect energy poverty in Hungary. This section analyses the policies relevant to this energy poverty. These include welfare policies, policies affecting energy affordability, and energy efficiency policies. We show that while the issue of energy prices has entered the political agenda, policy responses have so far failed to address energy poverty directly. They do not identify energy poverty, and nor do they offer effective measures to improve the situation of energy poor households. Finally, we briefly discuss how stricter EU regulations on heating devices impact the situation of energy poor people who rely on low-performing individual heaters.

Welfare policies

The welfare system in Hungary has undergone frequent changes in the past decades. Following the post-communist transition, it evolved into a hybrid, mixing pro-market and welfare-oriented principles and policies. The welfare system has been highly volatile as successive governments frequently undid their predecessors' social policies (Szikra and Tomka, 2009). Initial responses to the 2008 great financial crisis by the then-governing socialist government included drastic welfare cuts. And since 2010, when the right-wing conservative Orbán government took over, a comprehensive reshaping of the welfare system has been ongoing.

Welfare measures in this latter emerging system redirect support toward the better-off households while cutting benefits for vulnerable households. Central budget spending on social allowances was cut by 47 percent between 2010 and 2015 (Mózer et al., 2015). In the distribution of social benefits, meritocratic principles were introduced. For example, behavioral elements have been set by local municipalities as eligibility criteria for social allowances (Szikra, 2018a). For example, the length and level of unemployment benefit was drastically cut and linked to individual performance tests, and benefits activation policies for unemployed people were almost totally replaced by a compulsory public work program. While the program somewhat improved the income position of lower-income groups, it does not effectively lead individuals back to the primary labor market and makes affected low-income people vulnerable to arbitrary decisions of local elites (Koltai, 2017). In family policy, a reverse redistribution was introduced, including generous family tax credits available for employed parents only, freezing and thereby devaluing the amount of universal child allowance (Szikra, 2018b).

Regarding housing benefits, measures supporting housing affordability were reoriented in a way that increased the vulnerability of affected low-income households. From 2015 on, decisions on provision of support to ease affordability problems and to set eligibility criteria were placed at the discretion of local governments, contra the centrally regulated schemes applied in previous years. This resulted in a highly fragmented system of services, leaving more space for arbitrary decisions (Átol et al., 2016; Kopasz and Gábos, 2018; Kováts, 2015b). In addition, the availability of support dramatically worsened: housing benefits became unavailable in a quarter of Hungarian settlements (especially smaller settlements), the number of housing benefit recipients declined by 44 percent, and the number of sources allocated to such purposes in 2016 fell to just 55 percent of the 2014 budget allocation (Kopasz and Gábos, 2018).

Beyond reforms in the social welfare system, a radical tax reform was introduced between 2010 and 2013, including a flat tax rate. Childless members in the two highest income groups benefited from 74 percent in savings from the reform. The lowest income groups' tax burden slightly increased and families with more than three children in the lowest six income groups did not benefit from the reform (Tóth and Virovácz, 2013).

Polices related to energy affordability

As mentioned above, energy prices are traditionally a highly politicized issue in Hungary. During communism, energy prices were heavily subsidized to maintain a certain standard of living for households and to prevent social tensions which could threaten the political stability of the regime. Closely following the regime change, energy prices significantly increased. Though from the mid-90s this tendency became more moderate and fluctuating (during certain periods, even below the inflation rate), high energy prices relative to incomes contributed to the emergence and persistence of serious housing affordability problems. As a result, since the regime change, energy prices are a recurring topic on the political agenda. For most of the past three decades, policy-makers asserted control over energy prices through price regulation, which, according to governments' public communication, was needed to prevent households from experiencing higher energy prices (Energiaklub, 2016). Since the early 2000s, energy prices have been a recurring topic during election campaigns (Magyar, 2015).

In the 2002 election campaign, the then-governing right-wing Fidesz accused the opposition Socialist Party of intending to significantly increase gas prices, which the socialist party strongly rejected. This came after years of below-inflation increases in gas prices under the right-wing government, resulting in losses for gas providers and for the state budget (Gavra, 2005). The rejection of these price increases contributed to the Socialists' election success in 2002. However, in 2003 the government increased household energy prices, in order to limit providers' and state budget losses arising from a systematic difference between market prices and state-regulated household consumer prices (M. László, 2007). The same year, a general price subsidy for gas expenditures was introduced covering nearly all households. A more specific social targeting of the gas subsidy was instituted in 2007 for budget reasons (Fellegi and Fülöp, 2011).

In the run-up to the 2006 elections, the topic was again on the political agenda, voiced by then-opposition Fidesz to emphasize the Socialists' untrustworthiness, their lack of commitment to counteracting above-inflation energy price increases and people's decreasing living standards (M. László, 2007). Following Fidesz's landslide election win in 2010, a moratorium on gas and electricity price increases, and subsequently the reintroduction of state-regulated energy prices, was among the early political moves of the new Fidesz government.

The latest major political step concerning energy prices was the Utility Price Reduction Act mentioned above. This act was introduced by the Fidesz government in 2013–2014, when a series of legislative measures forced utility companies to reduce the prices of piped gas, electricity, district heating, water, and some other services, such as chimney sweeping, by 20–25 percent in total. Prices have been fixed at these levels up to the present. According to political analysts (Teczár, 2016; Török, 2013), the move was a reaction to the falling popularity of Fidesz and aimed at electoral gains in the 2014 Parliamentary elections.

Such considerations are supported by the emphasis the government has put on communicating measures to the public through various means. The new regulations obliged providers to highlight the amount of money saved due to the price cut using bright colors in the first lines of each energy bill issued to households. Energy price measures also became the key message in the 2014 election campaign, including a billboard campaign, a signature-collection campaign to support the measures, the assigning of a "Utility Commissioner" and organizing "Utility Forums" for public discussion of the new measures and the problems of energy prices. The measures and the subsequent campaign have been interpreted as an important factor in the re-election of Fidesz in 2014 (Böcskei, 2015). Even though the key political themes changed in the following years the government still used the Utility Price Reduction in its rhetoric in their 2018 Parliamentary election campaign – which was followed by re-election – claiming that opposition parties, if they won, would abolish these measures.

Also, in a first version of its National Energy and Climate Plan in 2010, the government considered the Utility Price Reduction as an effective response to energy poverty (Bouzarovski and Thomson, 2019). The Utility Price Reduction measures responded to a real social issue, as in the preceding decade the real value of energy prices increased by over 30 percent. Especially in the few years following the outbreak of the 2008 crisis, the burden of housing costs kept increasing. Housing costs peaked in 2010 at 25.2 percent of household income, and the share of energy expenditures within household expenditures peaked in 2011, at 17 percent on average.

However, energy and housing experts continue to criticize the energy price interventions of the Utility Price Reduction measure. The director of a leading Hungarian economic research company, GKI (Hegedűs, 2018), the director of a prominent think tank on housing and urban issues (Hegedüs and Péteri, 2015), and even the president of a research center very close to the government (Gyüre, 2014) pointed to several facts that modulate the effects of these interventions. They noted that vulnerable groups are systematically left out from some of the benefits of energy price reductions because these groups tend to use firewood heavily, which is not covered by the price cut. These groups also benefit less in absolute terms from the scheme because they consume less energy (Hegedüs and Péteri, 2015; Hegedűs, 2018). The favoring of higher income households, however, fits into a general tendency under the current government, which, instead of decreasing inequality, places support for high income families at the core of its social policy (Szikra, 2018b).

Other criticisms are that these interventions on prices are contradictory to EU regulations according to some authors (Hegedüs and Péteri, 2015) and that they are not feasible economically (Gyüre, 2014). Energy experts also warn that such price subsidies discourage energy efficiency investments even among those who can afford them. Finally, such measures do not only protect consumers from price increases in the global market; they also prevent possible savings when prices decrease, as consumer prices are fixed to the prices of a reference date in 2012.

The price cuts have substantially impacted energy suppliers. Faced with reduced incomes, suppliers have difficulties covering maintenance costs and realizing investments in infrastructure and in efficiency improvements. One consequence is that, since the launch of the Utility Price Reduction measures, the energy market in Hungary drastically changed: foreign companies have been leaving Hungary, while the government launched a national utility company. The latter covers an increasing proportion of the energy market, and includes the promise by the establishment of a non-profit entity for energy delivery

To conclude on energy affordability policies, note that the only socially targeted subsidy aimed at household energy spending is the "Social Fuel Wood Programme" introduced in 2011, which is available only for households living in villages with fewer than five thousand inhabitants. The budget for this program is less than 2 percent of the housing-related spending of the government. The system for disbursing this subsidy was heavily criticized, both due to its distribution being unequal and uneven and because it reinforced dependencies on local decision-makers (Bajomi, 2018). The average amount of subsidy per household is estimated to cover about 10 percent of actual heating costs in a typical house. However, in housing with poor energy efficiency the magnitude of support can be lower, so the scheme provides limited support for affected households (Bajomi, 2018). In certain cases, the quality of the fuel handed out in this program has been reported to be of low quality: this has been noted in the case of coal and of wet wood (Office of the Commissioner for Fundamental Rights, 2018).

Support for energy efficiency and housing

In Hungary, there currently are no predictable and accessible support schemes for energy efficient refurbishments. The only long-term and large-scale state-support programs for energy efficiency in the housing stock have been the so-called "panel programs" that aimed at renovating prefabricated panel buildings. Support from the panel programs was available between 2000 and 2009. Apartments in prefabricated buildings make up 18 percent of the housing stock, approximately 50 percent of which were at least partially retrofitted in the frame of such national programs (Ministry of National Development, 2015a, p. 31). Though prefabricated housing is widely considered as particularly exposed to energy poverty problems, households living in poorly insulated detached housing are more affected: according to previous research, 76–80 percent of energy poor households live in detached housing (Fülöp, 2011; Fülöp and Lehoczki-Krsjak, 2014) not covered by the panel program.

Currently, the only existing scheme for energy efficiency is the so-called "Warmth of Home" program, financed from incomes from the European Union Emission Trading Scheme. In it, grants are published for purchasing energy efficient devices, such as refrigerators and washing machines, as well as for energy efficiency investments in buildings. The subsidy contributed to dwelling insulation or to upgrading of district heating systems in 32 percent of the prefabricated buildings, with programs launched in 2008 and 2009. Further interventions between 2010 and 2017 increased energy efficiency in approximately forty-two thousand non-prefabricated apartments representing 0.9 percent of the total housing stock. The vast majority of these interventions were only partial retrofits (Ministry of Innovation and Technology, 2018). Access to funding was highly competitive, as the funds were often entirely allocated within a few hours of the opening of the application, which was available only online. The part of the program focusing on the retrofit of buildings or on upgrade of heating devices was virtually inaccessible to low-income households due the *ex post* payments of non-refundable funding, covering only 50 percent of the costs. A future opening of funding sources was announced for various dates, especially before elections but kept being postponed. This deteriorated the confidence of households in this instrument and decreased the predictability of the funding (Bajomi and Feldmár, 2018).

In the originally approved Environment and Energy Efficiency Operative Programme of Hungary for the 2014–2020 EU programming period, the government included an EU co-funded, non-repayable grant for the energy efficient refurbishment of dwellings. However, following the approval of the program by the European Commission, the government decided to use this source of funding for renovating public buildings instead of residential apartments, and the program was amended accordingly, despite criticisms by experts and NGOs (Weinhardt, 2018). Currently, the only ongoing renovation program for energy efficiency, financed by EU financial instruments, is provided through loans with a zero percent interest rate. However, even in this scheme, households must still pay significant initial administrative costs in addition to monthly payments and have to be creditworthy according to regular banking standards. Therefore, this is not suitable for low-income energy poor households living in low-efficiency housing.

In 2018, the state support of the long-existing contract savings schemes ("*lakástakarékpénztár*") was eliminated in a period of just three days. The scheme offered the possibility to save money on a monthly basis from four to ten years, with a 30 percent yearly state bonus after the payments, with an annual cap of 72,000 HUF, about 225 EUR. When the saving period ended, there was an opportunity to take up housing loans at a favorable rate to complete the amount needed for the purchase of a new home or the renovation of an existing actual one (Act on Home Savings Banks, 1996). The state support of home-saving funds existed for 22 years. Four million contracts were signed over this period and one-sixth of the population benefited from the support (Brückner, 2018). The savings, combined with the state subsidy, served the aim of renovations well, including amelioration of energy efficiency. The scheme also imposed less risk compared to market-based mortgages and it increased the monetary awareness of the population.

Even though it was used mostly by middle and higher status households and lower-income households were less likely be able to use it (Hegedüs, 2006), it was considered an important financial instrument supporting energy efficiency renovations. Since 2018, the state subsidy now gone, some providers have discontinued the scheme and the number of new saving contracts has significantly decreased. According to government communication, budget savings arising from the discontinuation of state subsidy for the scheme will be directed to the government's flagship housing program "Family Home Allowance" (Vass, 2018).

The Family Home Allowance scheme is a non-refundable state subsidy with the possibility of an additional state-subsidized loan. It is aimed primarily at supporting access to housing, preferably new housing and/or, in a specific version of the scheme, housing in certain rural areas, for households with children (with the maximum amount of subsidy and subsidized loan depending on the number of children). However, the scheme is not responding to the needs of funding for renovations. In its first two and a half years, only 0.4% of the subsidy contracted in the scheme was spent on renovation. In addition, certain provisions of the scheme, including a predefined time span of social security insurance as a precondition for reception, limited its availability for lower status households.

This leads to the conclusion that that there is currently a lack of policies and government programs in Hungary that could tackle energy poverty in a sustainable way.

Stricter EU regulations: improved standards or barriers to efficiency?

Several EU regulations, policies, and standards affect the use of energy and of related technologies in the member states. However, as the situation within each country can vary greatly, these measures can have drawbacks and backlashes. For example, when a household decides to change its heating system (whether gas or wood), the bureaucratic procedure and the cost of additional safety measures such as chimney refitting can discourage families from actually investing in the improvements, in particular when there is no adequate support for the provision of information or the additional financial burden this entails.

Also, the way in which certain EU targets are sometimes reached can have questionable results. The renewable energy share target set for Hungary for 2020, for example was reached with a change of reporting methodology in 2016, backdating the Eurostat SHARES data to 2010 (Mezősi et al., 2017). Several countries, including Hungary, changed their biomass accounting for the sales of solid biomass for households: official trading sales were replaced by self-reported usage based on national household surveys. As a result, Hungary reached the 13 percent renewable target in 2011 (European Commission, 2019).

As previously shown, solid fuels are predominantly used in inefficient, polluting ways. Therefore, the renewable energy target was reached by arguably one of the most polluting methods of energy production. In 2022, a new EU-wide standard regulating emissions from solid-fuel space heating will come into effect. This will most likely have a positive effect in wealthier countries; however, in countries like Hungary it is questionable how poor households will have access to improved technologies when cheaper – but more polluting – stoves will not be available anymore.

Conclusions

The aim of this chapter has been to examine how contextual and structural factors, as well as policy and politics, shape energy poverty in Hungary. The results offer some insights concerning the application of various energy poverty indicators to the Hungarian situation. It confirms that, despite the emerging scientific activity focusing on energy poverty in Hungary, there is no robust way yet to measure energy poverty in Hungary. There is a definite need for a robust set of indicators that makes the monitoring of energy poverty possible.

One important finding that emerged from this study is that, despite energy prices being positioned high up on the political agenda of political parties since the fall of communism, a consistent alleviation or prevention of energy poverty has not resulted. This is due to the lack of sufficient effort to ameliorate the building stock and to the fact that middle-class-oriented public policies in the fields of welfare, housing, and taxation have failed to address deeper social problems that are shaping energy poverty, in particular low incomes and unaffordable housing. Also, this study suggests that there is a persistent rural aspect of energy poverty, determined by a low-performing detached housing stock, dependence on solid fuels burnt in inefficient heaters, and an uneven distribution of incomes.

In the Hungarian context, the lack of predictable, accessible, and extensive public financing of energy efficient refurbishment on the one hand, and the burden of housing costs, lack of savings, and the previous negative experiences of disadvantageous loans on the other, are all factors that contribute to the low prospect of the refurbishment of the Hungarian building stock. Therefore, a large-scale, predictable renovation program that adequately includes dwellings of low-income households should be a key priority in the quest to alleviate energy poverty. In addition, other smaller scale interventions, such as energy advice, the education of social workers on energy poverty (identification, debt-management, etc.), an early warning system concerning households at risk of energy poverty, and improvements in the quality of the household appliances of the poorest households should also be envisaged to address the various aspects of energy poverty.

References

Act on Home Savings Banks, 1996. Article1, Act CXIII/1996. On Home Saving Banks. [1996. évi CXIII. Törvény a lakástakarékpénztárakról]. https://net.jogtar.hu/jogszabaly?docid=99600113.tv.

Átol, D., Kováts, B., Kőszeghy, L., 2016. Éves jelentés a lakhatási szegénységről. Habitat for Humanity Magyarország.

Bajomi, A.Z., 2018. A szociális tüzelőanyag-támogatás Magyarországon [The social fuel subsidy in Hungary]. Habitat for Humanity Hungary. https://habitat.hu/wp-content/uploads/2018/09/hfhh_tuzifa_tanulmany.pdf.

Bajomi, A.Z., Feldmár, N., 2018. Lakásminőség és energiaszegénység [Housing quality and energy poverty], in: Jelentés a lakhatási szegénységről 2018 [Report on housing poverty 2018]. Habitat for Humanity Hungary.

Balogi, A., Bódy, É., Kőszeghy, L., Koltai, L., Romhányi, B., Tomay, K., 2014. Éves jelentés a lakhatási szegénységről 2013. Habitat for Humanity Hungary.

Bernát, A., 2014. Leszakadóban: a romák társadalmi helyzete a mai Magyarországon, in: Társadalmi Riport 2014. TARKI Social Research Institute, Budapest, Hungary.

Böcskei B., 2015. Rezsicsökkentés: a közpolitikai változás mint politikai innováció. *Polit. Szle.* 24, 94–114.

Bouzarovski, S., Thomson, H., 2019. Transforming energy poverty policies in the European Union: second annual report of the European Union Energy Poverty Observatory, 55. www.energypoverty.eu/sites/default/files/downloads/observatory-documents/20-01/epov_pan-eu_report_2019_final.pdf.

Bouzarovski, S., Tirado Herrero, S., Petrova, S., Ürge-Vorsatz, D., 2016. Unpacking the spaces and politics of energy poverty: path-dependencies, deprivation and fuel switching in post-communist Hungary. *Local Environ.* 21, 1151–1170. https://doi.org/10.1080/13549839.2015.1075480.

Bouzarovski, S., Petrova, S., Tirado Herrero, S., 2014. From Fuel Poverty to Energy Vulnerability: The Importance of Services, Needs and Practices. *SSRN Electron. J.* https://doi.org/10.2139/ssrn.2743143.

Brückner, G., 2018. Lakástakarék: két nap alatt ledöftek egy iparágat [WWW Document]. https://index.hu/gazdasag/2018/10/16/ltp_ket_nap_alatt_ledoftek_az_iparagunkat/ (accessed 9.22.19).

Bucsky, P., 2018. Annyira pokoli lett egy egyszerű gázkazán-csere, hogy az emberek inkább áttérnek fafűtésre | G7 – Gazdasági sztorik érthetően. G7.hu.

Central Statistical Office, 2019a. Halálok: Túlságosan nagy hideg hatásai X31 [Cause of death: effects of extreme cold, X31]. http://statinfo.ksh.hu/Statinfo/QueryServlet?ha=NH3A03'.

Central Statistical Office, 2019b. 3.4. Fogyasztóiár-indexek (1960–) [WWW Document]. www.ksh.hu/docs/hun/xstadat/xstadat_hosszu/h_qsf001.html (accessed 2.16.20).

Central Statistical Office, 2019c. Energy 2006–2017 (3.1.2. Energia) [WWW Document]. www.ksh.hu/thm/3/indi3_1_2.html (accessed 3.26.19).

Central Statistical Office, 2018a. STADAT – 2.2.1.1. Az összes háztartás adatai jövedelmi tizedek (decilisek), régiók és a települések típusa szerint (2010–) [WWW Document]. www.ksh.hu/docs/hun/xstadat/xstadat_eves/i_zhc014b.html?down=1975.199951171875 (accessed 3.27.19).

Central Statistical Office, 2018b. 2.2.3.2. Lakással kapcsolatos adatok jövedelmi tizedek (decilisek), régiók és a települések típusa szerint (2011–) [WWW Document]. www.ksh.hu/docs/hun/xstadat/xstadat_eves/i_zhc019a.html (accessed 10.6.19).

Central Statistical Office, 2018c. Szociális Lakásgazdálkodás 2.7.7. (2004–2018) [WWW Document]. www.ksh.hu/thm/2/indi2_7_7.html (accessed 3.26.19).

Central Statistical Office, 2017. Piped gas supply. www.ksh.hu/docs/eng/xstadat/xstadat_annual/i_zrk003.html.

Central Statistical Office, 2016a. Non-occupied dwellings by property and number of rooms, floor area, year of construction, comfort, masonry and equipment. Microcenzus 2016. www.ksh.hu/mikrocenzus2016/docs/tablak/07/07_2_3.xls.

Central Statistical Office, 2016b. Population by nationality, highest educational attainment and age group, 2016. Microcenzus data. www.ksh.hu/mikrocenzus2016/docs/tables/02/02_00_national_1_2.xls.

Central Statistical Office, 2016c. A háztartások életszínvonala, 2016 [The living standard of households, 2016], 56. www.ksh.hu/docs/hun/xftp/idoszaki/hazteletszinv/hazteletszinv16.pdf.

Dobrescu, M., 2017. Romanian Forestry and Wood Products [WWW Document]. https://gain.fas.usda.gov/Recent%20GAIN%20Publications/Forestry%20and%20Wood%20Products%20_Bucharest_Romania_3-20-2017.pdf (accessed 9.21.19).

Ministry of Human Resources, 2015. A 2014–2020 közötti időszakra vonatkozó "A TELEPSZERŰ LAKHATÁS KEZELÉSÉT" megalapozó szakpolitikai stratégia. www.kormany.hu/download/3/a6/40000/Lakhat-Strat.pdf.

Energiaklub, 2016. Új lakásépítési forradalom – zöld forradalom nélkül? [WWW Document]. Energiaklub. /hirek/uj-lakasepitesi-forradalom-zold-forradalom-nelkul-3905 (accessed 10.8.19).

European Union Agency for Fundamental Rights, 2018. Combating child poverty: An issue of fundamental rights. Publications Office of the European Union. https://op.europa.eu/it/publication-detail/-/publication/331690aa-d1b6-11e8-9424-01aa75ed71a1.

Eurostat, 2020. Housing cost overburden rate. https://ec.europa.eu/eurostat/databrowser/view/tespm140/default/table?lang=en.

EU Energy Poverty Observatory, 2018. Arrears on utility bills by density of population [WWW Document]. EU Energy Poverty Obs. www.energypoverty.eu/indicator (accessed 3.3.20).

European Commission, 2019. Europe 2020 targets: statistics and indicators for Hungary. Share of renewable energy in gross final energy consumption [WWW Document]. https://europa.eu/webtools/rest/charts/export/html/ (accessed 10.7.19).

European Commission, 2018. Infringement – Internal energy market [WWW Document]. https://ec.europa.eu/commission/presscorner/detail/en/IP_18_4487 (accessed 2.9.20).

European Commission, 2015. Assessment implementation Hungary national strategy 2015. https://ec.europa.eu/info/sites/info/files/assessment_implementation_hungary_national_strategy_2015_en.pdf.

Eurostat, 2019. Gas prices for household consumers: Bi-annual data (from 2007 onwards) [nrg_pc_202]. https://appsso.eurostat.ec.europa.eu/nui/show.do?dataset=nrg_pc_202&lang=en.

Eurostat, 2018a. Inability to keep home adequately warm – EU-SILC survey [ilc_mdes01] [WWW Document]. https://appsso.eurostat.ec.europa.eu/nui/show.do?dataset=ilc_mdes01&lang=en (accessed 3.4.20).

Eurostat, 2018b. Arrears on utility bills – EU-SILC survey [ilc_mdes07] [WWW Document]. https://appsso.eurostat.ec.europa.eu/nui/submitViewTableAction.do (accessed 3.4.20).

Eurostat, 2018c. Total population living in a dwelling with a leaking roof, damp walls, floors or foundation, or rot in window frames or floor – EU-SILC survey ilc_mdho01 [WWW Document]. https://appsso.eurostat.ec.europa.eu/nui/show.do?dataset=ilc_mdho01&lang=en. (accessed 3.4.20).

Eurostat, 2018d. At-risk-of-poverty rate by poverty threshold and most frequent activity in the previous year – EU-SILC survey [ilc_li04] [WWW Document]. http://appsso.eurostat.ec.europa.eu/nui/show.do?dataset=ilc_li04&lang=en (accessed 4.7.19).

Eurostat, 2018e. Electricity prices for household consumers – bi-annual data (from 2007 onwards) [nrg_pc_204]. https://appsso.eurostat.ec.europa.eu/nui/show.do?dataset=nrg_pc_204&lang=en.

Fellegi, D., Fülöp, O., 2011. Szegénység vagy energiaszegénység? Energiaklub. https://energiaklub.hu/files/study/energiaklub_szegenyseg_vagy_energiaszegenyseg.pdf.

Ferge, Z., 1996. A rendszerváltás nyertesei és vesztesei, in: *Társadalmi Riport*. TÁRKI, Századvég, Budapest, pp. 414–443.

FRA and UNDP, 2012. the situation of Roma in 11 EU member states: survey results at a glance. Publications Office, Luxembourg.

Fülöp, O., 2011. Negajoule 2020. A magyar épületekben rejlő energiamegtakarítási lehetőségek.

Fülöp, O., Lehoczki-Krsjak, A., 2014. Energiaszegénység Magyarországon. *Statisztikai Szle*. 94, 12.

Gavra, G., 2005. Egy Gyulától lobbant be – Gázár és politika. Magy. Narancs.

Government of Hungary, 2019. Convergence program of Hungary 2019–2023. [WWW Document]. https://ec.europa.eu/info/sites/info/files/2019-european-semester-convergence-programme-hungary_hu.pdf (accessed 9.21.19).

Government of Hungary, 2018a. 39/2018. (III. 12.) Korm. rendelet a fűtési célú távhőszolgáltatást igénybe vevő lakossági felhasználók részére biztosított téli rezsicsökkentésről – Hatályos Jogszabályok Gyűjteménye [WWW Document]. https://net.jogtar.hu/jogszabaly?docid=A1800039.KOR&searchUrl=%2Fgyorskereso%3Fkeyword%3Dt%25C3%25A9li%2520rezsics%25C3%25B6kkent%25C3%25A9s%2520 (accessed 2.9.20).

Government of Hungary, 2018b. 1364/2018. (VII. 27.) Korm. határozat a téli rezsicsökkentés végrehajtását követően szükségessé váló további intézkedésekről – Hatályos Jogszabályok Gyűjteménye [WWW Document]. https://net.jogtar.hu/jogszabaly?docid=A18H1364.KOR&txtreferer=00000001.TXT (accessed 2.9.20).

Gyüre, J., 2014. Heim Péter: "Elmentünk a falig" [WWW Document]. Válasz.hu. http://valasz.hu/uzlet/elmentunk-a-falig-75595 (accessed 2.9.20).

Hegedüs, J., 2006. Lakáspolitika és a lakáspiac – a közpolitika korlátai. Esély 2006, 36.

Hegedüs, J., Somogyi, E., 2018. A lakások megfizethetősége és a társadalmi egyenlőtlenségek – a KSH 2015-ös lakásfelvétele alapján [Housing affordability and social inequalities – based on the 2015 housing survey of the Central Statistical Office], in: Miben élünk? Central Statistical Office, pp. 6–25.

Hegedüs, J., Péteri, G., 2015. Közszolgáltatási reformok és a helyi önkormányzatiság. *Szociol. Szle*. 2015, 30.

Hegedüs, J., Eszenyi, O., Somogyi, E., Teller, N., 2009. Lakhatási szükségletek Magyarországon. Városkutatás Kft.

Hegedűs, M., 2018. Méltányos rezsicsökkentést. GKI Economy Research Company. www.gki.hu/wp-content/uploads/gki/Hegedus_Miklos_Rezsicsokkentes_1.pdf.

Herrero, S.T., Ürge-Vorsatz, D., 2010. Fuel poverty in Hungary: a first assessment. http://3csep.ceu.hu/sites/default/files/field_attachment/project/node-3347/englishreportfuelpoverty.pdf.

Hungarian Central Statistical Office, 2011. Housing conditions: 2.22 Residential dwellings by age composition of resident, Census, 2011. [WWW Document]. www.ksh.hu/nepszamlalas/tablak_lakasviszonyok (accessed 4.7.19).

Hungarian Central Statistical Office, 2010. Magyarország 1989–2009: A változások tükrében [Hungary 1989–2009: In the light of change] [WWW Document]. www.ksh.hu/docs/hun/xftp/idoszaki/mo/mo1989_2009.pdf.

IEA (International Energy Agency), 2019. Heat generation by fuel (chart), Hungary. In: World Energy Balances. www.iea.org/statistics/?country=HUNGARY&year=2016&category=Heat&indicator=HeatGenByFuel&mode=chart&dataTable=ELECTRICITYANDHEAT.

Kasnyik, M., 2015. 10 grafikon, amit csak azok érthetnek meg, akik forintban kapják a fizetésüket [WWW Document]. kepek. http://kepek.444.hu/2015/02/05/ksh-arak-grafikonok (accessed 3.26.19).

Kemény, I., Janky, B., 2004. Roma population of Hungary 1971–2003, in: Roma of Hungary. New York.

Koltai, L., 2017. A közfoglalkoztatás hatása a helyi gazdaságra, helyi társadalomra. Hétfa Kutatóintézet, Budapest.

Koltai, L. (ed.), 2015. Lakhatási jelentés 2014 [WWW Document]. https://habitat.hu/wp-content/uploads/2018/08/HABITAT_2014_lakhatasi_jelentes_hosszu.pdf (accessed 10.20.19).

Koltai, L., 2013. Kihűlés és fagyhalál. Habitat for Humanity Hungary www.habitat.hu/files/Habitat_kihules2013.pdf.

Kopasz, M., Gábos, A., 2018. A szociális segélyezési rendszer 2015. márciusi átalakításának hatása a települési önkormányzatok magatartására, in: Kolosi, T., Tóth, I.G. (eds.), Társadalmi Riport 2018, pp. 328–349.

Kőszeghy (ed.), L., 2017. Lakhatási jelentés 2016.

Kőszeghy, L., Feldmár, N., 2019. Energiaszegénység Magyarországi kontextusban. www.elosztoprojekt.hu/wp-content/uploads/2019/12/LeaN%C3%B3ra.pdf.

Kováts, B., 2015a. Rezsitámogatás-csökkentés. Az új lakásfenntartási célú települési támogatások vizsgálata 31 önkormányzat példáján. *Esély*. 32.

Kováts, B., 2015b. Rezsitámogatás-csökkentés. Az új lakásfenntartási célú települési támogatások vizsgálata 31 önkormányzat példáján. *Esély*. 27, 29–60.

Lengyel, G., 2006. Cigánytelepek egykor és ma. Európai Összehasonlító Kisebbségkutatások Közalapítvány, Budapest.

Letenyi, L., Varga, A., 2011. Roma Társadalom., in: TÁMOP – 5.4.1–8/1 B Komponens. TeTT Consult Kft.

M. László F., 2007. Őszöd és a 2006-os kampányígéretek: Egymásra figyelve [WWW Document]. http://magyarnarancs.hu/belpol/oszod_es_a_2006-os_kampanyigeretek_egymasra_figyelve-67617 (accessed 3.4.20).

Magyar, L., 2015. A magyarországi lakossági villamosenergia-árak növekedésének okai az elmúlt két évtizedben. Published by Energiaklub. https://energiaklub.hu/files/study/energiaar_tanulmany_web.pdf.

MEKH (Hungarian Energy and Public Utility Regulatory Authority), 2019. Annual data – Hungarian energy and public utility regulatory authority [WWW Document]. www.mekh.hu/annual-data (accessed 10.6.19).

MEKH (Hungarian Energy and Public Utility Regulatory Authority) and PURA, 2019. Households with utility arrears, 2018. Data acquired by Habitat for Humanity Hungary. Available at https://docs.google.com/file/d/1Otpw6CGrcae-5COgBFzr1Xi1mHZDouEy/preview?rm=minimal.

MEKH and PURA, 2018. Közműhátralékkal rendelkező háztartások adatai, 2013–2017 [Data on households with utility arrears 2013–2017]. Data acquired by Habitat for Humanity Hungary. Available at https://docs.google.com/file/d/1Otpw6CGrcae-5COgBFzr1Xi1mHZDouEy/preview?rm=minimal.

Mezősi, A., Pató, Z., Szabó, L., 2017. Meg-megújuló statisztikák. REKK Found, 4.

Ministry of Innovation and Technology, 2018. Energiamegtakarítást eredményező pályázati programok adatai a 2018. Évi 22. Héten (támogatásban részesült pályázatok) [Data on programs resulting in energy savings in the 22nd week of 2018 (number applications that received support)]. In *Második Nemzeti Éghajlatváltozási Stratégia* [*Second National Climate Change Strategy*] (p. 69). https://nakfo.mbfsz.gov.hu/sites/default/files/files/N%C3%89S_Ogy%20%C3%A1ltal%20elfogadott.pdf.

Ministry of National Development, 2015a. National building energy performance strategy [WWW Document]. https://ec.europa.eu/energy/sites/ener/files/documents/2014_article4_hungary_en%20translation.pdf (accessed 4.7.19).

Ministry of National Development, 2015b. Hungary's national energy efficiency action plan until 2020 [WWW Document]. https://ec.europa.eu/energy/sites/ener/files/documents/hungaryActionPlan2014_en.pdf (accessed 10.6.19).

Misetics B., 2018. Sosem volt elég: lakásfenntartási támogatás 2015 ELŐTT. *Esély*, 2018/6, pp. 3–45. www.esely.org/kiadvanyok/2018_6/esely_2018-6_1-1_mitlesics_sosem_volt_eleg.pdf.

Misetics, B., 2013. Javaslatok egy egalitariánus lakáspolitikai reform fő irányaira: politikai és szakpolitikai vázlat. *Esély*, 2013/1. www.esely.org/kiadvanyok/2013_1/misetics.pdf.

Mózer, P., Tausz, K., Varga, A., 2015. A segélyezési rendszer változásai. *Esély*, 2015/3. http://esely.org/kiadvanyok/2015_3/2015-3_2-1_MozerTauszVarga_Segelyezesi_rendszer.pdf.

Office of the Commissioner for Fundamental Rights, 2018. Az alapvető jogok biztosa a szociális tűzifa támogatás elosztási gyakorlatáról – Közlemények – AJBH [WWW Document]. www.ajbh.hu/kozlemenyek/-/content/qzyKPkTyQAvM/az-alapveto-jogok-biztosa-a-szocialis-tuzifa-tamogatas-elosztasi-gyakorlatarol (accessed 10.6.19).

Pénzes, J., Tátrai, P., Pásztor, I.Z., 2018. Changes in the spatial distribution of the Roma Population in Hungary during the last decades [A roma népesség területi megoszlásának változása Magyarországon az elmúlt évtizedekben]. *Területi Stat.* 2018, 24.

Pósfai, Z., 2018. Annual report on housing poverty in Hungary, 2018 [WWW Document]. https://habitat.hu/wp-content/uploads/2018/11/HFHH_HousingPoverty_Report_2018_Final.pdf (accessed 3.27.19).

Reames, T.G., 2016. Targeting energy justice: exploring spatial, racial/ethnic and socioeconomic disparities in urban residential heating energy efficiency. *Energy Policy*. 97, 549–558. https://doi.org/10.1016/j.enpol.2016.07.048.

Szikra, D., 2018a. Távolodás ez európai szociális modelltől – a szegénység társadal ompolitikája. Magy. Tud. https://doi.org/10.1556/2065.179.2018.6.14.

Szikra, D., 2018b. Távolodás ez európai szociális modelltől – a szegénység társadalompolitikája. *Magy. Tud.* 179, 858–871. https://doi.org/10.1556/2065.179.2018.6.14.

Szikra, D., Tomka, B., 2009. Social policy in East Central Europe: major trends in the twentieth century, in: Cerami, A., Vanhuysse, P. (eds.), Post-communist welfare pathways: theorizing social policy transformations in Central and Eastern Europe. Palgrave Macmillan UK, London, pp. 17–34. https://doi.org/10.1057/9780230245808_2.

Teczár, S., 2016. "Gyakorlatilag nincs kormányzás" – Lakner Zoltán a rendszer természetéről, a Fidesz és az ellenzék esélyeiről. *Magy. Narancs.*

Teller, N., 2011. Adaptációs csapdák, in: Roma Kutatások 2010 – Élethelyzetek a Társadalom Peremén. Nemzeti Család- és Szociálpolitikai Intézet, Budapest, pp. 203–219.

Török, G., 2013. Török Gábor: Rezsi fel, rezsi le | Mandiner [WWW Document]. mandiner.hu. https://mandiner.hu/cikk/20130917_torok_gabor_rezsi_fel_rezsi_le (accessed 2.16.20).

Tóth, G.C., Virovácz, P., 2013. Nyertesek és vesztesek. Pénzügyi Szemle 385–400.

Tóth, I.G., Szelényi, I., 2018. Bezáródás és fluiditás a magyar társadalom szerkezetében, in: Társadalmi Report. Tárki Zrt., pp. 25–46.

Váradi, M., Virág, T., 2014. Faces and causes of Roma marginalization, in: Faces and causes of Roma marginalization in local contexts: Hungary, Romania, Serbia. Central European University, Budapest, p. 33.

Vass, Á., 2018. With or without a family plan it is tough to buy an apartment in Hungary. Hung. Today. URL https://hungarytoday.hu/housing-csok-ltp-family-plans-tough-buy-flat-in-hungary/ (accessed 2.16.20).

Weinhardt, A., 2018. Csendben beletörődött Brüsszel, hogy a családi házak helyett a középületekre szórjuk ki az EU-pénzt [WWW Document]. Portfolio.hu. www.portfolio.hu/unios-forrasok/20180626/csendben-beletorodott-brusszel-hogy-a-csaladi-hazak-helyett-a-kozepuletekre-szorjuk-ki-az-eu-penzt-290048 (accessed 10.18.19).

3 Inconsistencies in policy-making as drivers of energy poverty in Bulgaria

Maria Jeliazkova, Veneta Krasteva, and Douhomir Minev

Introduction

Energy poverty is a significant problem in Bulgaria (Hajdinjak and Asenova, 2019). A European Commission document in 2015 (European Commission, 2015, p. 42) depicts that Bulgaria has the highest share of population at risk of energy poverty, nearly three times higher than the EU average. Although some progress has been made in recent years – the share of the population unable to keep their homes adequately warm was almost two times higher in 2010 than now (Eurostat, 2020) – energy poverty in the country remains quite high. According to the European Survey on Incomes and Living Conditions (EU-SILC), in 2018, 33.7 percent of Bulgarian households were unable to keep their homes adequately warm in winter, while the EU average was 7.4 percent of households.

Following the general European guidelines on the Energy Union, Bulgaria implemented some policy measures to tackle its very high level of energy poverty. However, their practical implementation is influenced by a range of unresolved issues. We will address them in this chapter. Our goal is to assess the consistency of energy poverty policy-making in Bulgaria by looking at three of its characteristics: (1) the correspondence between definition, goals, and instruments, (2) the correspondence between programs and activities and (3) the correspondence across policy fields (Pal, 2014). After outlining the main dimensions of energy poverty in Bulgaria (section 1), we present our analytical framework (section 2). We then assess the consistency of energy poverty policy-making (section 3). The conclusion presents some recommendations for overcoming the inconsistencies in the current policy framework.

The multiple faces of energy poverty in Bulgaria

This section outlines the main characteristics of energy poverty in Bulgaria. We analyze four main dimensions: (1) the spread of energy poverty across the population, (2) the depth of energy poverty with respect to the structure of society, (3) the various health impacts of energy poverty, and (4) the low negotiating power of consumers to influence the factors that lead to energy poverty.

High levels of energy poverty

In Bulgaria energy poverty is widespread, as shown by various sources. A report published by the Centre for Energy Efficiency (EnEffect, 2019) states that about half of the Bulgarian population is on the verge of failing to cover their costs for normal, comfortable, healthy levels of heating. According to data from the EU-SILC survey, Bulgaria is ranked first in Europe on two energy poverty indicators: the extent to which people are unable heat and cool their homes comfortably in winter or summer – 33.7 percent of the Bulgarian population cannot afford to keep their home adequately warm (EU-SILC, data for 2018) and only 47.6 percent of the dwellings are comfortably cool during summer time, compared to the EU average of 79.1 percent (EPOV, 2012).

Regarding the affordability of energy, the National Statistical Institute (2018) estimates that, on average, expenses for housing, water, electricity, gas, and other fuels accounted for 14 percent of total household expenditures in 2018. And according to the European Energy Poverty Observatory (EPOV, 2018), 30.1 percent of the Bulgarian population had arrears on utility bills in 2018 – the second-highest figure in Europe, right after Greece (35.6 percent).

Energy poverty has been measured in Bulgaria by applying various methods that are used in the international literature. Based on data from the Household Budget Survey of the National Statistical Institute from 2014, energy poverty estimations range from 10 percent to 65 percent of the population (Peneva, 2019, p. 199). Ten percent of the population was estimated as energy poor based on the low-income high-cost method, about 35 percent based on the "poverty after fuel cost" method, 52 percent based on the "minimal income standard" method, and 64 percent using the "ten percent rule" method. It is obvious that the measurement of energy poverty depends highly on the indicators used and based on these data it could be suggested that at least 35 percent of the population in Bulgaria is energy poor.

Overall, these estimations suggest that energy poverty is widespread among the Bulgarian population. This situation can be related to three energy poverty drivers identified by Boardman (2010): low incomes, high energy prices, and low energy efficiency of homes. Regarding incomes, Bulgaria has the highest share of people at risk of poverty or social exclusion in the EU, at 32.8 percent of the population (Eurostat, 2020 – data for 2018).

Regarding energy prices, their continual rise, which is disproportionately high relative to the rise in household incomes, affects the purchasing power of Bulgarian households. Electricity prices for households in Bulgaria are stated to be the lowest in the EU. However, when compared with the incomes and purchasing power of the population, energy prices are rather high. As stated by the Economic and Social Council (2015, p. 13):

> The dynamics of energy prices has outpaced the dynamics of basic earnings. ... For the period 2007–2014 the general price level (CPI) has increased by 26.1%, with a significant contribution to this growth with

> rates of price increase of five energy groups, and some of them significantly outpacing the general price level. The group of fossil fuels for example increased by 38.3%; gaseous fuels by 39.1%; liquid fuels for household purposes by 35.6%; heating by 29.4%; and electricity by 20.7%.

This undoubtedly contributes to a further increase in the overall level of living costs in the country.

Finally, the poor quality and low energy efficiency of homes create problems too. Bulgarian housing stock is old, prefabricated panel homes make up a large share of the stock, and certain groups of the population are faced with poor living conditions (Economic and Social Council, 2015, p. 14). The latest census of the population, in 2011, showed that over 20 percent of the homes in the country are panel buildings, and approximately 10 percent of them are urgently in need of renovation of the interior installations and hydro-insulation of the roofs, of heat insulation of the external walls, and a change of windows and doors (Ministry of Energy, 2017). And the National Housing Strategy for 2018–2030 identifies major problems: "Poor housing maintenance ... low energy efficiency, low comfort and unhealthy living conditions" (Ministry of Regional Development and Public Works, 2018, p. 16).

While renovation programs of multi-family homes have been implemented, these do not seem to include plans to change the heating system, or to enable the use of biomass or natural gas (Adjarova, 2016). Due to the low purchasing power of households, adding expenses to fund energy efficiency is not possible for most people. As indicated in the National Housing Strategy, most homeowners in Bulgaria would need public support to be able to renovate their homes (Ministry of Regional Development and Public Works, 2018, p. 17).

Deep energy poverty pockets

Energy poverty is very high among households in the four poorest deciles of the population, who spend almost one-fifth of their budget on housing, water, electricity, gas, and other fuels (Table 3.1). Their budget share increased between 2009 and 2018, while the budget share for the whole population has decreased, showing an increase of inequalities in the Bulgarian population. This is visible also in the evolution of the income share of electricity, gas, and other fuels between 2005 and 2015, with a marked increase in this share for the three lowest income deciles (Table 3.2).

Despite high budget shares, most of these households save heavily on energy costs and need to make choices, for example between food, medicine, and heating. Some researchers have pointed out that household energy consumption is relatively low, which "can be partly explained by the desire of individuals to reduce thermal energy costs and the high levels of energy poverty in the country, which results in under heating/cooling and poor

Table 3.1 Share of expenses for housing, water, electricity, gas, and other fuels in the structure of the total households' expenditure by income decile groups (percent)

Year	*Average*	*Decile 1*	*Decile 2*	*Decile 3*	*Decile 4*
2009	14.5	13.7	15.4	16.7	16.7
2018	13.6	18.3	18.6	17.1	17.2

Source: Compiled by the authors based on data from the Household Budget Survey, National Bulgarian Statistical Institute, 2018 and 2009.

Table 3.2 Consumption expenditure for electricity, gas, and other fuels as a share of income by income quintile (percent)

	Income quintile 1	*Income quintile 2*	*Income quintile 3*	*Income quintile 4*	*Income quintile 5*
2005	9.9	10.6	9.7	9.4	8.0
2015	14.2	12.2	11.0	9.4	7.8

Source: Compiled by the authors based on data from the European Energy Poverty Observatory based on Eurostat data, "Structure of consumption expenditure by income quintile and COICOP consumption purpose" (code: hbs_str_t223).

living conditions" (iBRoad, 2017, p. 2). This issue of low energy consumption requires particular attention in the Bulgarian context: identifying and measuring it would be helpful to assess so-called "hidden" energy poverty that is not captured by analyses that focus on a high budget share of energy.

Energy poverty is particularly prevalent among families of retirees, families with children, and households with unemployed and economically inactive members (Zahariev et al., 2016, pp. 16–17). Energy affordability problems have been highlighted by data from the Bulgarian Household Debt Survey conducted by the Institute of Sociology in 2010. That data reveals that some households, especially retirees and Roma people, resort to loans for paying utility bills (for water, electricity, and heating) (Institute of Sociology, 2010, pp. 147–149). The interest rates of these loans are generally high, resulting in a loan trap for households, with a high risk of non-repayment and resulting over-indebtedness.

Still, among the above-mentioned programs for building renovations, there are no programs specifically targeting people in deep energy poverty.

Various health impacts of energy poverty

The health impacts of energy poverty take various forms. First, the combination of under-heating in winter, of arrears on utility bills, and of the need to borrow money to pay these bills result in a vicious circle: health and well-being are impacted, resulting in an increase in healthcare costs and often in the need to choose between medicines and heating.

Second, high levels of energy poverty are often associated with cold-related mortality. According to EPOV data, the share of excess winter mortality in 2014 in Bulgaria was 16.9 percent.[1] Cold-related deaths are regularly reported in the media. Victims are usually either elderly or homeless people. In the winter of 2018, two new cases of cold-related mortality were reported in the media (Newspaper 2018). In the first case, an 80-year-old man lived in a house without electricity and he used the timber floor of the house for heating. In the second case, an 85-year-old widow, after her home burned down in a fire, moved to live in an outbuilding that had neither electricity nor water. The media reports on these cases have resulted in the topic being publicly debated. However, there have been no visible discussions among policy-makers about energy poverty in the country.

Finally, another health impact of energy poverty is air pollution: air pollution levels are reported to be dangerously high in Bulgaria and the country "has the highest rate of premature deaths due to air pollution in the EU (217 deaths per 100,000 people; 13,000 premature deaths per year)" (Heilmann et al., 2019, p. 20). The main sources of air pollution include solid fuels, such as coal and wood, used for heating (Heilmann et al., 2019). Here, it is important to note that the usage of solid fuels, the type of heating equipment, and the fact that households live in buildings with low energy efficiency are neither the consequences of households' preferences nor caused by of a lack of information. They are due to the very low purchasing power of incomes, which does not allow money to be allocated to cleaner fuels and cars, nor to improving the housing stock. Moreover, switching to cleaner fuels may simply be impossible in certain areas because of the lack of access to alternatives.

Low negotiating power of consumers to influence the factors leading to energy poverty

Under the conditions of widespread and deep energy poverty in Bulgaria, it is important to consider the possible ways in which citizens could have an impact in trying to find solutions. For example, citizens could be involved in developing a vision to overcome their difficulties. However, their ability to influence political decision-making on rights linked to services of general interest (especially the right to energy) are extremely limited.

The market for residential electricity and gas customers is regulated by the Energy and Water Regulatory Commission (EWRC). Typically, EWRC, which is in charge of price regulation, justifies price increases arguing that energy distribution/supply companies ask for much higher increases than those actually enacted, without further proof of the need to increase prices, the profits of the energy companies, and the mechanisms for consumer feedback on decisions taken. This situation has already resulted in serious public protests; for example, in 2013, after another major price increase, people reacted by setting themselves on fire (Reuters, 2013; Novinite.com, 2013; OffNews, 2013, February 26) and began civil protests that led to the fall of the government.

A common element to all these instances is the poor negotiating power of citizens and the ability of energy producers, distributors, and suppliers to dictate conditions. This is further reinforced by the regional monopoly of the electricity distribution companies.[2] An example showing the asymmetry of power between distributors/suppliers and consumers is the implementation of a decision of the Supreme Court. This decision obliges one of the electricity distribution/supply companies operating in southeast Bulgaria (OffNews, 2013, August 13) not to collect a fee when restoring electricity after a disconnection. However, the company, along with other companies operating in Bulgaria, continued to collect the fee based on the argument that the court's decision did not cancel the "inclusion" fee (that is, the fee to pay for reconnection) as the company has two different arms – one for the supply of energy and another one for the distribution, the latter being responsible for the network through which the electricity reaches the houses. According to the company, the court's decision only cancelled some of the clauses of the supplying company, but not those of the distributing one. According to the Ombudsman of Bulgaria, the law states clearly: "If the reason for the disconnection disappears, the obligation is to supply electricity and no fees are due" (Nova TV, 2015). Thus, although the Bulgarian ombudsman receives more than three hundred complaints yearly against this practice of electricity suppliers, the companies found a legal opportunity and continue to collect a fee for restoring electricity after a disconnection.

These examples illustrate the low leverage of consumers in relation to the providers or state authorities. This situation prevents consumers from being able to affect any necessary reforms related to clean energy and the environment. It also contributes to increased consumer vulnerability, to a lack of attention of policy-makers to situations of deep energy poverty, and to distrust by citizens in political decisions and in the regulators. Frequently, solutions are sought through conflicts, outbursts, and clashes, rather than through calm, well-intentioned, and prudent discussions of consistent steps.

Having outlined the main dimensions of energy poverty in Bulgaria, we now propose to investigate the policies that address them. Our focus will be on the consistency of energy poverty policy measures. We first outline our analytical framework and then we analyze policy consistency on different dimensions.

An analytical framework to assess policy consistency

Policy consistency seems to have attracted scientific attention in different fields, especially in the analysis of EU external activities (see for instance: Cremona, 2008; Gauttier, 2004; Krenzler and Schneider, 1997). Following

Krenzler and Schneider (1997, p. 134), policy consistency can be defined as "coordinated behaviour based on agreements where comparable and compatible methods are used in pursuit of a single objective and result in an uncontradictory policy." This consistency implies: an absence of contradictions, rules of delimitation by effective allocation of tasks between actors and instruments, avoidance of duplications and gaps in institutional competences, and synergies between norms, actors, and instruments (Minkova, 2011).

As pointed out by Lieu et al. (2018, p.1):

> Particular challenges that policy makers face when designing policies to solve complex problems are that policy instruments (PI) do not exist in isolation; they are instead part of a wider context, interacting with other PIs through the responses of targeted stakeholders. In some policy areas, such as energy and climate policy, interactions will become even more important as an increasing diversity of stakeholders, including citizens, need to be involved. Although these aspects may be acknowledged, there are limited methods applied that account for complex policy and human interactions in the policy making and implementation process.

Our aim is to analyze energy poverty policies in Bulgaria and to assess how policy consistency affects the state and prospects of energy poverty. In this regard, the analytical framework proposed by Pal (2014), which looks at three types of consistency, appears suitable for discussing energy poverty in Bulgaria:

> Policies are expected to be consistent in several interrelated ways … First … we expect policies to have an *internal consistency* among the three elements of problem definition, goals, and instruments. Second, we expect a policy to have *vertical consistency* in the sense that the programs and activities undertaken in its name are logically related to it … A third type of consistency is *horizontal consistency*, or consistency within the wider policy space and across policy fields, not just within them. This is an expectation that what governments do in one field will not contradict what they do in another … This type of consistency is important in democratic politics, since it implies that there is an underlying philosophy of government that cuts across all policy fields.
>
> (Pal, 2014, pp. 13–14)

In the next section, we discuss energy poverty policies within the abovementioned three types of policy consistencies: internal, vertical, and horizontal, in turn. Accordingly, it is important to clarify the scale that could be used for assessing the level of consistency. Lieu et al. (2018) propose a three-level scale to evaluate consistency in environmental policy mixes. The

Table 3.3 Extent of policy consistency

Level	*Description*		
	Internal consistency – quality of correspondence between definition, goals, instruments	*Vertical consistency – quality of correspondence between programs and activities*	*Horizontal consistency – quality of correspondence across policy fields*
Very high	High and meaningful correspondence		
High	Good correspondence		
Moderate	Relative correspondence		
Low = inconsistency	Questionable quality of correspondence		
Very low = high inconsistency	Mismatch		

Source: Authors' analysis.

scale differentiates among an inconsistent policy mix, a weak consistency of policy instrument mix, and moderate to strong consistency of policy instrument mix. It assesses the extent of consistency within a policy mix based on pair-wise policy instrument interactions. Table 3.3 provides a summary of the approach.

Using this scale, we will assess the level of correspondence in policies addressing energy poverty in Bulgaria. Additionally, we will also take advantage of the well-known policy cycle framework (agenda setting – policy formulation – decision-making, implementation – policy evaluation – back to agenda setting) which covers the whole policy-making process and allows identification of the stages that most need improvements (Bridgman and Davis, 2003; Howlett and Ramesh, 2003; Werner and Wegnich, 2007). We consider that combining the two frameworks, of policy consistency and of the policy cycle, is helpful to better identify the major problem areas in the case of energy poverty policy in Bulgaria.

Assessing public policies' (in)consistencies in Bulgaria

Having defined our analytical framework, we now discuss the level of correspondence for each of the three types of policy consistency – internal, vertical, and horizontal – in Bulgaria, in reference to energy poverty.

Correspondence between problem definition, goals, and instruments

Internal consistency refers to the correspondence between problem definition, goals, and instruments. In the terminology of the policy cycle, this type of consistency is directly related to the first stage of that cycle, agenda setting, which includes problem definition.

In Bulgaria, anti-poverty policies are the responsibility of the Ministry of Labour and Social Policy (MLSP). The ministry elaborates and adopts anti-poverty strategies, and plans for their implementation, coordinates various measures, and reports results and challenges. However, there is no definition of energy poverty in the country, despite the accumulating recommendations to develop such a definition – in research analyses (Jeliazkova, 2013) and from official positions (Economic and Social Council, 2015). In fact, despite the growing recognition of the problem of energy poverty, even the term itself is not used in key strategic documents. Instead, the "National Strategy against Poverty and for Promoting Social Inclusion 2020" written in 2013 only mentions "targeted heating support" and reports how much public money is provided under this program (Council of Ministers of the Republic of Bulgaria, 2013).

The Strategic Plan 2017–2021 adopted by MLSP (Ministry of Labour and Social Policy, 2020), which implements the governmental program for the period, presents reported and planned targeted heating assistance by number of persons and families. The plan supports around 220,000 households, with a target of 250,000 households. However as already mentioned, about one-third of Bulgarian households, which is about 1 million households, cannot keep their home warm in winter. Thus, the plan's aim is to cover only about a quarter of those in need. The arguments for this goal are not explicit, but they are usually based on claims that these are the available funds. Moreover, the freedom of the MLSP to take decisions in the field is severely limited by its subordinate role to the Ministry of Finance, which is in a higher hierarchical position than other ministries.

The absence of a definition of energy poverty both impedes a clear identification of the problem and conveniently reduces the need for effective policy efforts in the field (Kulinska, 2017; Peneva, 2014; 2019). The MLSP program has no formulated goal to address energy poverty. As B. Petkov, the Minister of Labour and Social Policy, pointed out in 2018:

> The Ministry of Labour and Social Policy position on energy poverty is that it is a complex issue that should not be solved only with targeted heating support, even more so that there is no clear definition … Financial support is only a means of blunting the severity of poverty, but not a means of escaping poverty.
>
> (Statement before the National Assembly of the Republic of Bulgaria, Parliamentary Control, 2018)

The aim of the heating support is to provide ongoing income support for certain households for the winter season. The level of support provided for a household depends on a number of conditions (such as age, family status, property status, etc.). It is not enshrined in law and is decided for each year on the basis of an ordinance that has been amended many times since 2008. Its level is in no way linked to the official poverty line. It is a conditional means-tested benefit available to people who can demonstrate that their income is below a certain level and apply for it.

In more detail, heating support is provided to households with income below a certain threshold, the so-called "minimum guaranteed income." In the period 2008–2018, the minimum guaranteed income was BGN 65 per month (33.23 euro) and from 2019 is BGN 75 per month (38.35 euro) (Council of Ministers, 2018). It is further differentiated according to age, family status, disability, etc. and thus it has a sanctioning effect on people of working age. Also, the threshold is bound by conditions regarding property ownership. The eligibility conditions have been further complicated in 2019 by the introduction of a new concept of "Basic Heating Income," which is double the amount of the differentiated "Guaranteed Minimum Income." A sanction has been added for children not attending school and their parents, thus making heating support conditional on school attendance. Receiving energy assistance requires claimants to present an application and local social workers to implement a social test, aimed to determine if the person meets the necessary conditions. If the claimants are found to be eligible for energy assistance, then they receive aid determined for the year in question, depending on prices.

The consistency of the approach for heating assistance can be analyzed as follows: a consistent policy approach (very high, high, or relative) would mean that the problem is clearly defined, the goals are set following this definition, and instruments are in correspondence and aim to reach the stated goals. In Bulgaria, it seems that the approach is upside down, based on instruments that implicitly define goals and the problem, which undermines policy effectiveness. Based on available funds, the characteristics of the assistance provided are determined instrumentally, and the goals, objectives, and definition of the problem are implicitly set out in the instruments. There is no doubt that such a picture can hardly be characterized as particularly consistent. The problem is not clearly defined, the objectives are minimized, and the instruments developed and implemented are based on other, external considerations. Certainly, such an instrumental approach is unable to organize effectively to overcome the wide and deep energy poverty in Bulgaria. Based on this, the internal consistency could be characterized as low.

Correspondence between programs and the activities undertaken

In our analytical framework, the correspondence between programs and activities undertaken is called *vertical consistency*. In the policy cycle, questions related to vertical consistency are linked to the implementation stage and include "street bureaucracy": social workers in the local social assistance offices implement the program, assessing whether the applicants are eligible and, if so, granting them heating assistance.

In Bulgaria, the social assistance system is highly centralized. Local social assistance offices and the social workers in them have a low degree of discretion in decision-making. It can be assumed that they usually follow the detailed requirements for obtaining heating support. However, the heating support system, as well as the refusals and complaints related to it, have not been independently evaluated and interactions between different stakeholders have not been analyzed. The lack of such data and analyses makes it difficult to assess the implementation of the program. Although fragmentary and incomplete, we have used information from the website of the MLSP,[3] where citizens can ask questions about social assistance and expect answers from the central administration. Based on the content analysis of this publicly available information (questions and answers,[4] each with a name, date, and time) the following conclusions can be drawn. (1) In general, the questions have been answered quickly – usually either the next day or within the same day. (2) Case-specific questions were the most common. This suggests that, for many citizens, the complex and often changing conditions/requirements for access to heating support remain unclear. It also seems that on-site social workers often do not provide consultancies on a case-by-case basis, but bureaucratically refer to regulatory documents (i.e., they do not consider the concrete case in order to give a direct answer but point out only the part of the law/normative document where the claimant can look for an answer). Thus, for claimants it remains unclear whether they are entitled to heating benefits based on their income. (3) The questions on the website outline problems mainly linked to the eligibility conditions.[5] (4) Most often, the answers send people to the local office in their place of residence. Only in some cases is some advice provided.[6] (5) Unlike other types of social assistance, cases of corruption are less frequently reported for heating support.[7] (6) Sometimes questions concern operational issues related, for example, to the timely provision of forms.[8]

The brief overview of the implementation of programs can be summarized as follows. To a large extent, activities at local social assistance offices seem to strictly follow the program requirements. In the vast majority of cases, claimants receive heating support if and to the extent that they meet the changing conditions for eligibility. In borderline cases (for example, when incomes are quite near the threshold), citizens do not seem well informed about whether they are eligible or not to receive the heating support. Qualitative consultations in specific cases are most probably lacking. People are expected to apply for heating allowance, which is usually time-, effort-, and money-consuming and often linked to different bureaucratic obstacles, especially when people with mobility difficulties – disabled or elderly people – are concerned. It can be assumed that the quality of the services provided by the social workers is different from case to case, not only as a result of their competences and experience, but also depending on the locality where they work. In deprived communities, where the flow of applications is higher, the quality is likely to be lower. In terms of evaluating program implementation processes, independent evaluations are probably needed, as well as an analysis of rejected applications. The issue of understanding why qualified households do not apply for heating assistance, which is not at all the focus of research or social assistance, also deserves further attention.

Based on this review, we can evaluate the degree of vertical consistency of energy poverty policies in Bulgaria. A very high or high degree of vertical consistency would mean that activities strictly follow the programs. Local social workers would implement the defined programs coherently and on time, providing quality advice for people concerned. The link between programs and activities would be improving, based on independent assessments and lessons learned. Our analysis suggests, rather, that the degree of vertical consistency is moderate. Activities follow the programs, but gaps exist in relation to border cases and to the quality of social work. And independent assessments of the implementation of programs are underestimated or missing.

Consistency across different policy fields

Analyzing the consistency across different policy fields allows assessing its *horizontal consistency*. The present section analyzes some of the policy fields that have the closest connection to energy poverty, such as environmental policy, energy policy, energy markets, and welfare policies.

Environmental policy

Bulgaria has many problems in terms of environmental quality. Air pollution in cities is mainly due to the use of solid fuels for heating and old equipment. At the same time, measures to move toward lower carbon emissions and/or domestically procured sources of energy supply (renewables) have driven up energy prices for households and have changed the nature of energy provision in structural terms, sometimes leading to more environmental damage when, for example, as already mentioned, electricity heating is replaced by wood and coal heating. Protests by the population against electricity price increases in the previous years have brought an end to the inclusion of new renewables in the system. In this context, it is not surprising that the idea of energy prosumers is still missing in the public debate in Bulgaria.

However, strategic documents related to climate change (Sixth[9] and Seventh[10] National Communication on Climate Change) do not discuss the interrelation between the climate change mitigation measures undertaken and energy poverty. Some environmental assessments of climate policy (Za Zemiata, 2019, p. 41) argue that poverty is used as an argument to avoid undertaking the necessary environmental measures, for example, when supporting heating with wood and coal, both of which are highly polluting of the environment. In fact, the two policy areas (clean environment policy and anti-energy poverty policy) block each other in a vicious circle.

Energy policy

Energy policy and the diversification of energy sources are constant topics in the political debate in Bulgaria, but these topics are mainly approached from a geopolitical perspective. In general, the official debates on energy policy do not focus on how to provide cheaper energy for the population, or on how to deal with social problems in the energy field.

Bulgaria follows the privatization and deregulation policies pursued throughout Europe during the past two decades. If energy policies in Bulgaria have some specificity, it is related to the relatively high level of corrupt practices in the implementation of energy sector liberalization (Dimitrov and Yosifova, 2017; Pashev et al., 2006; Stefanov et al., 2014). Signs of this could be found in privatization deals, in the signing of long-term contracts for compulsory purchase of output, payment of "non-produced output," various fees and other instruments through which producers' risks and losses are passed on to customers. In the end, these practices result in prices which are too high for consumer incomes, and perhaps in disproportionately high profits for some energy companies (Investor.bg, 2019; News.bg, 2016). Such peculiarities shape the energy production, distribution, and supply as a rent-seeking activity. And what can be seen as particularly worrying in relation to energy poverty is not so much the lack of a functioning energy market, but the doubts that the envisaged measures (creating market actors and liberalization) will be able to generate an efficient energy market.

Energy "market"?

The Ministry of Energy and the EWRC in Bulgaria are coordinating the liberalization of energy markets. To this end, several changes have been made to the Energy Act, which was last amended in October 2019: all electricity producers with capacities between 1 and 4 megawatts are included in the energy exchange system as of July 1, 2019. The aim is to improve energy pricing (market prices instead of regulated prices) by increasing the supply of electricity and fostering competition. However, according to representatives of Bulgarian trade unions, expectations regarding the effects of liberalization are too optimistic. Some experts (Grigorova, 2016) state that the consequences, especially for the most vulnerable people, will not be positive as "after full liberalization of the energy market: 1.5 million Bulgarians will not be able to pay their electricity bills."

As a result of the steps undertaken to create a free market after 2016, a free and a regulated electricity market already exists in Bulgaria. Some consumers buy electricity from the regulated market, at prices set by the EWRC, while others purchase energy from different suppliers at freely negotiated prices. The presence of the two types of markets makes it possible to compare them. Generally, it would be expected that the creation of an energy exchange would have a beneficial effect on the energy market in the country, as more transparency is achieved through public auctions (everyone sees the prices and quantities of energy sold or purchased) and this creates conditions for competition, therefore maintaining adequate electricity price levels. However, the effects of the newly created market do not meet these expectations, which can be attributed to different causes.

1. *Opportunities for free market manipulation.* In 2018, employers' organizations and EWRC simultaneously advised small businesses, which were integrated into the free market, to return to the regulated market (Kirkova, 2018), after 892 applications had been submitted by customers seeking return to the regulated market in western Bulgaria. In particular, nationally representative employers' organizations (Bulgarian Chamber of Commerce, the Association of Industrial Capital, the Bulgarian Chamber of Commerce and Industry, and the Chamber of Employers and Industrialists) stated in a letter to their members that participation in the free market is unprofitable and advised fifty thousand companies (affiliated with low voltage networks) to move from the free to the regulated market.

According to observers, the reason is the speculative rise in electricity prices – between 35 percent and 65 percent – by traders in the free market, where prices reached BGN 90–100 per megawatt hour (MWh), while on the regulated market prices were BGN 76 per MWh. Electricity prices provoked a lively debate in July 2019, reaching an average price of BGN 195.40 per MWh, and in peak times BGN 246.72 per MWh. Thus, electricity for businesses became more expensive than average European levels. The Association of Employers' Organizations argues that certain companies manipulate electricity trading by buying cheap energy from state-owned power plants (through manipulated auctions) and then selling it expensively. The director of the National Electric Company[11] confirmed the opinion of the employers, saying that there were "scandals" and "perversions" in the functioning of the market.

The Association of Employers' Organizations in Bulgaria also points out that there are many cases of drastic increases in electricity prices. In a letter to the Minister of Energy and disseminated also through the media, the Association states: "Systematically since the beginning of 2019 … the average price of electricity for the industry in Bulgaria is higher than the one in Germany, France, Austria, Czech Republic, Slovakia, Switzerland" (Association of Bulgarian Employers, 2020). At a subsequent meeting, an agreement was reached on the need to develop a draft legislative change in order to implement adequate measures.

2. *Legacies.* In addition to speculation, another important reason for the increase in electricity prices has been mentioned by the Association of Bulgarian employers: the surcharge for "preferential prices" at which electricity is purchased from some power plants. This surcharge is a specific component of electricity prices, one unique to Bulgaria among European countries. This is a legacy of the period of intensive privatization ("preferential prices" for the purchase of electricity were regulated during the privatization of power plants) and contributes significantly to "Bulgarian enterprises buying the most expensive electricity in Europe," as the Association of Bulgarian Employers (2020) claims.

3. *Informal networks and corruption.* Informal networks and their rent-seeking practices are particularly powerful tools for distorting the market mechanism (Newspaper "Sega," 2019). District Heating Sofia is a municipal company and a large electricity producer. According to the legislation (after changes in the legislation from July 1, 2018) the company is obliged to sell its electricity on the market. But instead of selling through the market, the company directly signed a contract with Grand Energy Distribution Ltd for the sale and purchase of electricity. Grand Energy Distribution Ltd was selected without competition and, in addition to the opportunity to make profits from the resale of the electricity, it also received a commission for the services it provided. The contract was concluded in accordance with Art. 100, § 6 of the Energy Act, which allows power plants to sell their energy

through the so-called "balancing group coordinator," a trader who supplies energy from different places (Newspaper "Capital," 2019).

According to analyses of this development, District Heating Sofia sells its electricity to Grand Energy Distribution Ltd at BGN 71 per MWh, only slightly above the market minimum determined by the regulator (EWRC). This is not illegal, but it gives the private trader a real opportunity to make additional profits from the market. As a result of the deal, District Heating Sofia has missed out over BGN 13 million, which have been transformed into profits for Grand Energy Distribution.

Thus, it can be concluded that the establishment of free market institutions does not guarantee efficient market mechanisms resulting in adequate electricity prices through the "free play of market forces." The effectiveness of a "good" market mechanism is not just about building market institutions and liberalizing the market but depends strongly on the whole environment in which the market mechanisms operate. Ideally, this environment has many components – for example, a regulatory framework that does not contain contradictory norms or "loopholes" that allow for abuse; effective institutions for implementing this framework and accountability mechanisms; market actors who are motivated to follow the regulatory framework and do not purposely seek rents by exploiting weaknesses in the regulatory framework and weaknesses of institutions; and adequate policies that stimulate market actors toward adequate (pro-market) behavior and do not create contradictory effects that are mutually neutralized. In the absence of such a favorable environment, market mechanisms fail or can be manipulated by market participants. It is therefore interesting to monitor the liberalization process and its effects – what kind of liberalization contributes to increasing or reducing energy poverty, how liberalization interacts with the context (normative, economic, social) in which it is applied, and how that context influences liberalization and its effects. All these are especially important for a country like Bulgaria, the poorest in the EU.

Welfare policies

The social structure in Bulgaria favors the accumulation of poverty. The country displays a very high level of inequality in primary income distribution (in terms of the Gini index before social transfers, Bulgaria ranks third after Ireland and Latvia for the highest inequalities in EU) and very high inequalities in redistribution (in terms of the Gini index after social transfers, Bulgaria ranks first in the EU, having the highest level of inequalities). Additionally, the country has the lowest share of social benefits in the national gross domestic product compared to all other countries in the EU. The taxation system based on a flat tax and regressive tax wedge[12] transfers the costs for a clean environment to average-income and low-income groups.

Thus, in summary, the overall effect of the pair-wise interactions of the different policy fields is as follows:

Table 3.4 Effects of the pair-wise interactions of different policy fields

	Interplay with energy poverty
Environmental policy	Environmental policy and anti-energy-poverty policy mutually paralyze each other
Energy policy	Increases in energy prices result in energy poverty increase
Liberalization, privatization, and energy market	Liberalization, privatization, and the development of the energy market have resulted in further energy price increases leading to higher levels of energy poverty
Welfare policies	Welfare policies supporting low quality of jobs, including pressure on average and low incomes and growing income inequalities, result in support and increase of energy poverty

Source: Authors' analysis.

High consistency across different policy fields would mean that different policy fields interact positively with each other, mutually reinforce each other, and produce synergies, leading to higher effectiveness than the sum of effects of individual instruments in the mix, and/or higher cost-effectiveness. What we witness in Bulgaria in reference to its energy poverty policy approach is the opposite: a mismatch between different policy fields that often contradict each other and mutually erode and paralyze each other. On this basis, our analysis shows that the horizontal policy consistency with respect to alleviating energy poverty is very low.

Conclusion and policy recommendations

Our analysis outlines the low internal consistency (a low correspondence between definition, goals, and instruments) and the very low horizontal consistency of public policies (a low correspondence between different policy fields) in Bulgaria with respect to energy poverty. The main problems lie in the stages of agenda setting and in the lack of evaluations that could allow for the clear identification of the problems. This suggests that the current policy approach is unable to contribute effectively to decrease the rate of energy poverty in the country.

If the horizontal inconsistency between different policies persists, this could have various effects. On the one hand, the energy poverty support system could become more and more expensive, trying to catch up with price increases. On the other hand, pollution generated by solid fuels could increase. Therefore, different policies need to be coherent and integrated. Efforts to achieve consistency across policies are an important condition for overcoming the problem of energy poverty. In this regard it is very important for public policies to consider the well-being of people and clean energy together and to clearly define possible steps and timing so that people's welfare is not compromised while pursuing climate change goals. A possible tool for this could be to establish and enforce an adequate minimum income as well as a progressive taxation system and a special investment fund to cover the costs of energy efficiency improvements.

Toward ensuring this consistency, it would also be essential to increase the participation of citizens, including vulnerable citizens, in decision-making mechanisms and in seeking appropriate scenarios, rather than pursuing the same policies linked to narrow economic and fiscal interests.

Finally, we recommend that evidence-based solid evaluations, including participatory action research, be implemented in the policy cycle to provide reliable assessments of policy effectiveness and to contribute to improving policies.

Notes

1 Excess winter deaths are defined as the ratio of deaths in winter (December to March) to those in other months.
2 ENERGO-PRO took over the electricity distribution in northeast Bulgaria in 2012; CEZ Bulgaria AD is the company responsible for maintaining the network and distribution in western Bulgaria. Electricity Distribution South AD is the electricity distribution company in southeastern Bulgaria. The three energy distribution companies have also set up separate energy supply companies.
3 www.mlsp.government.bg/index.php?section=FA2&fcat=17, accessed 7 March 2020.
4 A total of 104 questions along with answers to each of them are available. All available questions and answers are analyzed.
5 Excerpts from publicly available citizens' questions addressed to the Disability, Equal Opportunities and Social Assistance Directorate at the MLSP: (1) My grandmother is 91 years old living with my dad and they are both retired, my grandmother's pension is 250 BGN (around 128 euro). Could she successfully apply for heating aid? They both meet all other requirements; (2) I would like to ask why my mother was receiving heating assistance for two years, and this year she was refused assistance, and was told that she should receive a pension up to 275 BGN (around 141 euro) monthly. She is 75 years old. She receives the minimum pension of about 210 BGN (around 107 euro) monthly, as well as widow's pension from my father of about 97 BGN (around 49 euro) monthly; (3) My mother is 64% disabled and 68 years old, and my father is 70. How the heating support is calculated and what is its amount?; (4) In the area where I live, there is an elderly woman 85 years old. She lived in a home owned by her and her deceased spouse. Currently she lives at this home together with her daughter and her family of four people and she has a pension of 250 BGN (around 128 euro). From the local Social Assistance Office they have refused her heating assistance. Could you please answer me if the refusal is justified?; (5) I would like to ask if it is appropriate to ask my mother where her children live and, who is her personal doctor when applying for electricity support?
6 Excerpts from publicly available answers to citizens' questions addressed to the Disability, Equal Opportunities and Social Assistance Directorate at the MLSP: Please note that the assessment of compliance with all statutory conditions and

requirements can only be made on the basis of the results of the social test conducted by the social workers of the Social Assistance Directorate at your parents' address. If your testimony is confirmed by the social test, your parents will be able to get help ... We recommend that your parents submit an application to the Social Assistance Directorate at their current address; We note that it is not the amount of the pension but the amount of income earned for the relevant period that is important in determining the right to heating support.

7 Excerpt from the website of the MLSP: Question (Q): Hello, this lady (name deleted) has received social assistance from (local office deleted) for heating even though the whole of Bulgaria saw her profit of BGN 100,000 (around 51,129 euro) last December. But why should we be surprised when the neighborhood says that if you give 50 BGN (around 26 euro) you will get the support? Answer (A): The Social Assistance Agency has been notified of the necessity to check.

8 Excerpts from the website of the MLSP: (1) (Q): Why in (name of the city deleted), are the forms for applying for heating support not available, while on the website of the Social Assistance Agency it is declared that applications could be submitted from 01.07.19? Is it only valid for Sofia (the capital)? (A): This is an organizational matter within the competence of the Social Assistance Agency. Please ask there; (2) (Q): A relative of mine has lost the order for heating (voucher), which is in kind. Is there an opportunity for him to receive a duplicate? The Social Assistance Agency replies that if a voucher is damaged, destroyed or lost, a duplicate cannot be issued. Is this really so? Please do not refer me to the Social Assistance Agency. (A): The order is the basis for issuing an invoice, which creates an obligation to pay from the Social Assistance Agency to the merchant-issuer of the invoice. Therefore, no duplicate or substitute document could be issued.

9 Sixth National Communication on Climate Change, Republic of Bulgaria, 2013, https://unfccc.int/sites/default/files/vi_nc_bulgaria_2013_22102014_final_-_resubmission.pdf.

10 Seventh National Communication on Climate Change, Republic of Bulgaria, 2018, www4.unfccc.int/sites/SubmissionsStaging/NationalReports/Documents/274195_Bulgaria-BR3-NC7-1-VII_NC_Bulgaria_2018.pdf.

11 Statement given in media: "On the electricity exchange: twice as expensive than in EU," www.mediapool.bg/ (in Bulgarian).

12 "Sum of personal income tax and employee plus employer social security contributions together with any payroll tax less cash transfers, expressed as a percentage of labour costs." – OECD, Glossary of Statistical Terms, https://stats.oecd.org/glossary/detail.asp?ID=7273, downloaded on 25.03.2020.

References

Adjarova, L. (2016). "Good European practices for – the REACH example." Paper presented at XII National Conference of Bulgarian Energy Agencies and Centres, 06.12.2016, Sofia (in Bulgarian).

Association of Bulgarian Employers. (2020). "Declaration AOBE on the price of electricity for industry in Bulgaria," https://aobe.bg/category/news/.

Boardman, B. (2010). *Fixing Fuel Poverty: Challenges and Solutions*. Earthscan, London.

Bridgman, P. and Davis, G. (2003). *What Use Is a Policy Cycle? Plenty, if the Aim Is Clear*. National Council of the Institute of Public Administration, Australia.

Cremona, M. (2008). "Coherence through law: What difference will the Treaty of Lisbon make?" *Hamburg Review of Social Sciences*, vol. 3, no. 1, pp. 11–36.

Council of Ministers. (2018). Decree No. 305 of 19 December 2017 determining new monthly amount of the guaranteed minimum income. www.mlsp.government.bg/uploads/1/zakoni/pms-305-20171.pdf.

Council of Ministers of the Republic of Bulgaria. (2013). "National strategy against poverty and for promoting social inclusion 2020," www.strategy.bg/StrategicDocuments/View.aspx?lang=bg-BG&Id=790.

Dimitrov, N. and Yosifova, M. (2017). "Corruption in energy sector." In: *Science, Education, Security*. New Bulgarian University, Sofia (in Bulgarian).

Economic and Social Council. (2015). "Measures to overcome energy poverty in Bulgaria." Work, Income, Living Standards and Industrial Relations Commission, and Commission for Social Policy, www.parliament.bg/pub/ (in Bulgarian).

EnEffect. (2019). "Centre for Energy Efficiency, report: Energy poverty in the light of local elections: What can the state and municipalities do to solve the problem of high heating bills?" www.eneffect.bg/ (in Bulgarian).

Energy Act. https://lex.bg/laws/ldoc/2135475623 (in Bulgarian).

EPOV. (2012). Dwelling comfortably cool during summer time. EU Energy Poverty Observatory, EU-SILC data. www.energypoverty.eu/indicators-data

EPOV. (2014). Excess winter mortality/deaths. EU Energy Poverty Observatory, EU-SILK data. www.energypoverty.eu/indicators-data.

EPOV. (2018). Arrears on utility bills. EU Energy Poverty Observatory, EU-SILC data. www.energypoverty.eu/indicators-data.

European Commission. (2015). "Commission staff working document, monitoring progress towards the Energy Union objectives – concept and first analysis of key indicators," SWD (2015) 243, 18.11.2015, https://eur-lex.europa.eu/legal-content/EN/TXT/PDF/?uri=CELEX:52015SC0243&from=EN.

Eurostat. (2020). "Inability to keep home adequately warm – EU-SILC survey, last updated 27.03.2020," https://appsso.eurostat.ec.europa.eu/nui/show.do?dataset=ilc_mdes01&lang=en accessed 30.03.2020.

Gauttier, P. (2004). "Horizontal coherence and the external competences of the European Union," *European Law Journal*, vol. 10, no. 1, pp. 23–41.

Grigorova, V. (2016). "After full liberalization of the energy market: 1.5 million Bulgarians will not be able to pay their electricity bills," Bulgaria ON AIR TV, 08.06.2016. www.bgonair.bg/.

Hajdinjak, M. and Asenova, D. (2019). "Sustainable energy consumption and energy poverty: challenges and trends in Bulgaria," in Fahy F., Goggins G. and Jensen C. (eds) *Energy Demand Challenges in Europe*. Palgrave Pivot, Cham.

Heilmann, F., Reitzenstein, A. and Amon, A. (2019). "The political economy of the low carbon transition, climate and energy snapshot: Bulgaria," www.euki.de/wp-content/uploads/2019/02/Bulgaria-PEMM-Snapshot.pdf.

Howlett, M. and Ramesh, M. (2003). *Studying Public Policy: Policy Cycles and Policy Subsystems*. Oxford University Press, Toronto.

iBRoad. (2017). "Factsheet: Bulgaria, current use of energy performance certificates and potential links to iBRoad," http://ibroad-project.eu/wp-content/uploads/2018/01/iBROAD_CountryFactsheet_BULGARIA.pdf.

Institute of Sociology. (2010). *The Financial Crisis and the Debt of the Bulgarian*. Chance Publishing House, Sofia (in Bulgarian).

Investor.bg. (07.01.2019) "Most energy companies in Bulgaria report growth in profit for the first nine months of 2018," www.investor.bg/energetika/472/a/povecheto-energiini-drujestva-v-bylgariia-otchitat-ryst-v-pechalbata-za-pyrvite-devet-meseca-na-2018-g-274746/ (in Bulgarian).

Jeliazkova, M. (2013). "Energy poverty in EC – state, researches and policies," in *Energy Poverty*. Union of Economists in Bulgaria (in Bulgarian).

Kirkova, T. (2018). "Small business escapes the free electricity market," *Policy*, 12 October.

Krenzler, H. and Schneider, H. (1997). "The question of consistency," in Regelsberger, E., Schoutheete de Tervarent, P. and W. Wessels. (eds) *Foreign Policy of the European Union: From EPC to CFSP and beyond*. Lynne Rienners, London.

Kulinska, E. (2017). "Defining energy poverty in implementing energy efficiency policy in Bulgaria." *Economic Alternatives*, issue 4, pp. 671–684.

Lieu, J., Spyridaki, N., Alvarez-Tinoco, R., van der Gaast, W., Tuerk, A. and van Vliet, O. (2018). "Evaluating consistency in environmental policy mixes through policy, stakeholder, and contextual interactions," *Sustainability*, vol. 10, no. 6, pp. 1–26.

Ministry of Energy. (2017). "National long-term programme for promoting investments for fulfillment of the measures for improving the energy characteristics of buildings from the public and private national residential and commercial housing fund 2016–2020," https://ec.europa.eu/energy/sites/ener/files/documents/bg_building_renov_2017_annex_6_of_neeap_bg.pdf (in Bulgarian).

Ministry of Labour and Social Policy. (2020). "Updated strategic plan 2021," www.mlsp.government.bg/ (in Bulgarian).

Ministry of Regional Development and Public Works. (2018). "Housing strategy 2018–2030," www.strategy.bg/PublicConsultations/View.aspx?lang=bg-BG&Id=3607 (in Bulgarian).

Minkova, V. (2011). "Horizontal consistency in the European Union external action," Munich, GRIN Verlag, www.grin.com/document/179295.

National Statistical Institute. (2018). Household expenditure. www.nsi.bg.

News.bg. (25.04.2016) "Our energy companies are increasing their profits," https://news.bg/energy/energiynite-ni-druzhestva-uvelichavat-pechalbite-si.html, last accessed 30.03.2020 (in Bulgarian).

Newspaper. (01.12.2018) "Cold and neglect kill single elderly people in Knezha," *24 hours*, www.24chasa.bg/novini/article/7179926 (in Bulgarian).

Newspaper. (21.06.2019) "Who loves 'District Heating – Sofia.'" *Capital*.

Newspaper. (03.08.2019) "Employers refer to the Commission on Protection of Competition for a cartel in the electricity market," *Sega*, www.segabg.com/hot/rabotodatelite-sezirat-kzk-za-kartel-na-pazara-za-elektroenegriya (in Bulgarian).

Nova TV. (11.08.2015) "No switch-on fee if our electricity is disconnected due to unpaid electricity bills," https://nova.bg/news/view/2015/08/11/ (in Bulgarian).

Novinite.com. (22.02.2013) "Varna man setting himself on fire turns Bulgarian Jan Palach," www.novinite.com/articles/148124/Varna+Man+Setting+Himself+on+Fire+Turns+Bulgarian+Jan+Palach.

OffNews. (26.02.2013) "A third person sets himself on fire in despair," https://offnews.bg/balgaria/treti-chovek-se-samozapali-ot-otchaianie-164803.html (in Bulgarian).

OffNews (13.08.2013) "Power reconnection fee is prohibited," https://offnews.bg/kakvi-sa-pravata-ni-kato-potrebiteli/zabraniavat-taksata-za-povtorno-vkliuchvane-na-toka-230741.html (in Bulgarian).

Pal, L. (2014). *Beyond Policy Analysis – Public Issue Management in Turbulent Times.* Nelson Education, Toronto.

Pashev, K., Djulgerov, A. and Kaschiev, G. (2006). "Corruption in public procurements: risks and counteraction." Center for the Study of Democracy, Sofia (in Bulgarian).

Peneva, T. (2014). "Energy poverty: the Bulgarian case." International Association for Energy Economics, www.iaee.org.

Peneva, T. (2019). "Energy poverty in Bulgaria," in *Yearbook of the Faculty of Economics and Business Administration*, Sofia University "St. Kliment Ohridski," vol. 1, pp. 187–210 (in Bulgarian).

Reuters. (06.03.2013) "Bulgaria mayor resigns after man sets himself on fire," www.reuters.com/article/uk-bulgaria-government/bulgaria-mayor-resigns-after-man-sets-himself-on-fire-idUKBRE9250ME20130306.

Statement before the National Assembly of the Republic of Bulgaria, Parliamentary Control. (5/10/2018). B. Petkov. www.parliament.bg/bg/parliamentarycontrol/ID/720.

Stefanov, R., Galev, T., Tsanov, M., Vladimirov, M. and Yordanova, M. (2014). "Energy sector governance and energy (in)security in Bulgaria." Center for the Study of Democracy, Sofia.

Werner, J., Wegnich, K. (2007). "Theories of the policy cycle," in Fischer, F., Miller, G.J., Sidney, M.S. (eds) *Handbook of Public Policy Analysis*. CRC Press, Boca Raton.

Zahariev, B., Grigorova, V. and Yordanov, I. (2016). *Energy Poverty in Bulgaria.* Open Society, Sofia (in Bulgarian).

"Za Zemiata" (Friends of the Earth Bulgaria). (2019) "Comments on the Integrated Energy and Climate Plan (IPEC) of the Ministry of Energy," www.zazemiata.org/wp-content/uploads/2019/05/ZaZemiata_stanovishte_INPEK_april_2019.pdf, (in Bulgarian).

4 Slovenia

The role of NGOs in the emergence of energy poverty as a policy issue

Lidija Živčič and Tomislav Tkalec

Introduction

Energy poverty in Slovenia is manifested due to a combination of factors that are often cited when describing the phenomenon: the triad of low incomes, high energy prices, and poor energy efficiency of homes. However, their mechanisms differ somewhat from those generally present in other post-socialist countries in the region. Slovenia has experienced rising income inequalities, but to a lesser extent than other countries in the region. Energy prices have also increased at a slower pace following a smooth process of energy market liberalization. Nonetheless, the strongest cause of energy poverty is the poor quality of the housing stock, due to which such poverty remains rather hidden as households typically apply energy rationing behavior, often preferring to give up on some of their basic needs in order to be able to enjoy a minimum standard of thermal comfort.

In Slovenia the first report on energy poverty was elaborated by the national government in 2010. Yet, the first actions to tackle the phenomenon have been taken by nongovernmental organizations (NGOs). After having been engaged extensively at the grassroots level while implementing European grant projects on energy poverty, NGOs learned that Slovenia was in need of more overarching and comprehensive approaches to energy poverty. They persuaded the government to adopt national level policies and programs to combat the phenomenon based on realities on the ground. NGOs worked extensively to identify typical characteristics of households affected by energy poverty along with the most appropriate interventions and methodologies to alleviate this poverty. They also engaged actors across sectors in dialogue on the topic as the complexity of the issues demanded many points of view. These efforts resulted in the first structural approaches to energy poverty, including a national program for advising households affected by energy poverty. Slowly the topic developed into one that became an increasingly consistent element on the political agenda and in a variety of projects. Today there are several support programs available to households affected by energy poverty. This experience shows that cooperation between NGOs and the government can be very fruitful, although challenging. The following chapter describes this interaction from the perspective of one of the NGOs involved in the process, namely FOCUS Association for Sustainable Development.

Energy poverty in numbers

After the disintegration of the former Yugoslavia, Slovenia, which compared to all the other break-away countries was better off due to its softer form of socialism, embarked upon a process of transition from socialism to capitalism. This transition was also somewhat milder for Slovenia. Unlike other countries in the region, Slovenia did not experience rapid privatization or market liberalization. This proved to be quite important in the context of energy poverty for two key reasons. First, energy prices did rise along the way, but not as dramatically as was the case in other countries, allowing Slovenian society a more gradual adjustment. Second, there was less susceptibility to social disparities. Slovenia is one of the OECD countries with the lowest Gini coefficient, which means that income inequality is rather low (Figure 4.1). Given the situation, experts link energy poverty mainly to the poor state of buildings (Tkalec and Živčič, 2014) (Stegnar et al., 2018).

Despite a typically fortuitous regional positioning, the picture looks different with a European eye proving that real convergence in terms of energy well-being for the population exists. The Energy Poverty Index shows that Slovenia is more affected by the phenomenon than the EU28 average: the index in Slovenia is 13.9, while the average index across the EU28 is 10.8. The Combined Energy Poverty Index for Slovenia (16) is also higher than the EU average (13) (Figure 4.2) (OECD, 2019). There are various explanations for this. According to a recent study, the share of the Slovenian population living in dwellings with leaking roofs, damp walls, floors or foundation, or rotten window frames is among the highest in the EU, as is the share of households late with payments for utility services (Primc et al., 2018). The European Energy Poverty Index also shows that the explanations lie with high energy expenditures as a share of total household expenditure due to the poor state of the dwellings, which renders families unable to keep their houses cool in summer and warm in winter. More recently, the question of summer energy poverty has been raised. Due to the inability of low-income households to keep their dwellings cool in summer, Slovenia counts among the countries that are under visible stress of summer heat (The Right to Energy Coalition, 2019).

In terms of policy responses, the same study shows that Slovenia ranks number 15 in the EU with regards to domestic energy poverty alleviation efforts (with 62.05 points); the European Domestic Energy Poverty Index ranking is 20 (51.34 points) and the European transport energy poverty index is 14 (74.99 points) (OpenExp, 2019).

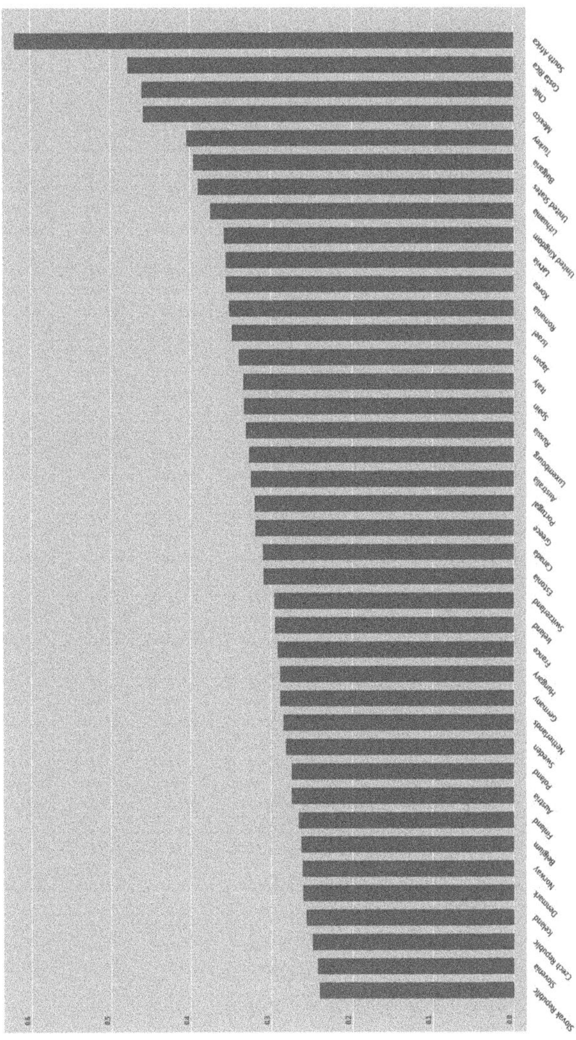

Figure 4.1 Comparison of Gini coefficients in OECD countries.

Source: OECD, 2019 (used with permission).

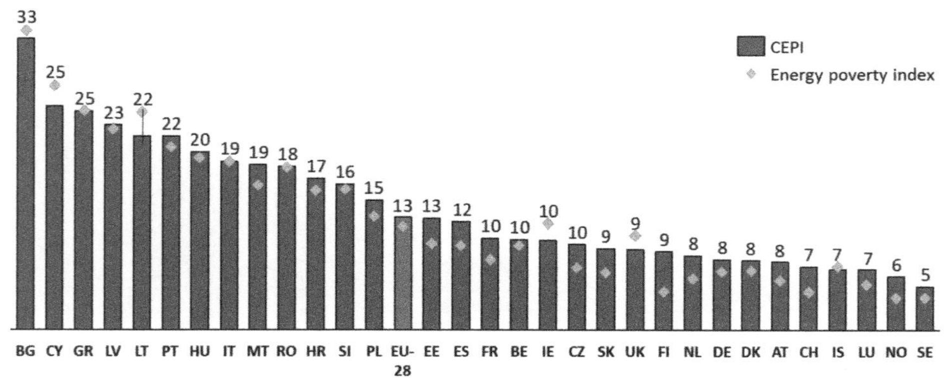

Figure 4.2 Comparison of CEPI and Energy Poverty Index for EU28.

Source: Primc et al., 2018 (reprinted from Maxim et al. 2016) (used with permission).

Thus, in Slovenia energy poverty is mainly linked to the poor state of the buildings, whereas all the other factors count to a lesser extent. As a result, instead of paying lip service to measures that target income or prices, nongovernmental actors have identified in practice structural interventions to generate the greatest impact on energy poor households.

In Slovenia, as in other countries of the region, energy poverty is closely related to general poverty. The lack of financial means pushes many households into a situation of limitation when it comes to energy well-being. However, many of those that are just above the "at risk of poverty" threshold are also profoundly affected. Generally speaking, statistics and practice have shown that the most vulnerable categories are single-parent households (single mothers especially) and the elderly (especially those who are single) (Tkalec and Živčič, 2014).

The latest study on energy poverty in Slovenia shows that households in the lowest income quintile spend more than 17 percent of their disposable income on energy (Primc et al., 2018). That share is rising because the increase in income lags behind increases in energy costs. This plays out differently for different population segments. For example, in 2016, 5 percent of low-income households (i.e., those with incomes below 60 percent of the median income) were unable to heat their homes adequately, whereas the EU28 average was above 11 percent. Based on this indicator, the overall situation in Slovenia is clearly better than the European one. That means that more low-income households can afford adequate heating in Slovenia. For low-income single-member households, the 2016 average for Slovenia was also lower than the EU28 average, albeit only slightly (Slovenia 18.5 percent, EU28 19.3 percent); most notably, the challenge for low-income single-member households to adequately heat is markedly higher than for other low-income categories. The same is valid for low-income households with one senior adult (Slovenia 16.1 percent and EU28 17.4 percent, both in 2016). On the contrary, for households with one adult under the age of 65 within this income category, the proportion of those unable to afford adequate heating is already higher than the EU28 average (Slovenia 20.9 percent, EU28 20.6 percent, in 2016). So, despite the favorable comparison to the EU average, some subgroups of low-income households are particularly worse off, including single-member households, elderly people, and some of the working poor (Primc et al., 2018). And as explored below, the situation is less encouraging if dependent children come into question.

People affected by energy poverty tend to feel stigmatized and typically mistrust institutions (Tkalec and Živčič, 2014). Hence, just a small proportion even seek help. This is why cases of energy poverty remain rather silent and are not easy to pinpoint. Moreover, the categories identified above as being particularly vulnerable have no uniform spatial distribution. This means that measures for eradicating energy poverty cannot be targeted to specific geographical areas or neighborhoods, complicating the process of identification and the types of actions that that can be applied (Živčič et al., 2016). Therefore, more in-depth knowledge of the factors and characteristics is needed.

Energy prices are also pertinent to energy poverty. It is common knowledge that Slovenians spend a relatively higher proportion of their income on energy than other Europeans (Slovenia 5.4 percent, the EU28 3.9 percent, in 2017) (Primc et al., 2018). One of the reasons for this is that energy products for a typical Slovenian household customer, adjusted for purchasing power, and in particular natural gas and fuel oil, are more expensive than the EU average, whereas electricity is at the level of European prices (Primc et al., 2018). Energy prices have been rising for the past ten years, but not significantly, with the exception of heating oil. This situation is only to a small extent due to the price of commodity on the global market, and is more importantly due to a government strategy intended to decrease the use of heating oil in favor of more sustainable fuels. Between 1999 and 2011 there was a gradual increase in heating oil prices from 0.2 EUR/liter to 0.9 EUR/liter (Statistični urad Republike Slovenije, 2012). As a result, consumption of heating oil in households decreased by 53 percent between 2008 and 2015, while the use of wood-based fuels increased proportionally; in 2015, wood fuels represented 57 percent of all energy used for space heating.

Except for those in the more vulnerable categories, households have adjusted to the rising prices relatively well by switching fuels. Table 4.1 shows that between 2002 and 2015 the share of energy expenditure in proportion to available income increased especially for the first income quintile and to a lesser extent for the second and third quintile, whereas for households in the fourth quintile the share of expenditure decreased. In the fifth quintile the share of energy expenditure remained quite stable (Primc et al., 2018). The most likely reason for this situation is that for the lower quintiles, income is growing more slowly than energy prices. What is more, households in the lower quintiles are less able to switch fuels when prices increase, especially if switching involves investments in new technologies, such as new boilers, which poor households most likely cannot afford. On the contrary, they apply daily coping strategies, including forgoing some of their other basic needs in order to secure a minimal level of thermal comfort. Also, while performing household visits to energy poor homes, energy consultants identified the problem of self-disconnection from district heating networks – a manifestation of the trade-off between energy services and other basic needs (Robic et al., 2017).

From an income–price relation perspective, energy poverty is also linked to the late payment of utility bills. In Slovenia utility bills cover the costs for electricity, heating, water, and waste. According to this indicator, Slovenia ranks among the countries with the highest share of late payments for household expenses in EU28. Between 2013 and 2016, the overall share with late payments decreased from 19.7 percent to 15.9 percent, but among the low-income group it remained as high as one-third. Comparatively, in

Table 4.1 Share of the available resources of households spent on energy in dwellings by income quintiles (percent)

Income category	*2002*	*2008*	*2015*	*Difference between 2008 and 2015 (absolute %)*
1st quintile (lowest)	13.1	15.1	17.7	2.6
2nd quintile	9.3	10.2	10.1	–0.1
3rd quintile	7.2	7.9	7.7	–0.2
4th quintile	6.1	6.0	5.9	–0.1
5th quintile (highest)	4.4	4.0	4.0	0.0

Source: Primc et al., 2018.

EU28 18.5 percent of the low-income groups cannot afford to pay their bills. The situation is particularly worrisome for those low-income households consisting of one adult and dependent children. While for this category the EU28 average for late payments is around 23 percent, in Slovenia the share is around 60 percent (Primc et al., 2018).

But beyond income and price-related explanations for energy vulnerability, the situation is perpetuated by the generally poor state of the buildings which require high volumes of fuel to heat. The proportion of the population living in dwellings with leaking roofs, damp walls, foundations or floors, or broken window frames is among the highest in the EU. In 2016, 24 percent of Slovenians lived in dwellings with such problems. The situation is much worse for the low-income category, where almost 40 percent live in houses with these conditions. This is around 9 percentage points more than the EU28 average for the same category. Single-parent households with dependent children stand out as a particularly affected subgroup. If the single parent is the mother, the percentage rises to 44 percent, which is 18 percentage points higher than the EU28 average for the parallel group (Primc et al., 2018). The poor quality of housing stock in Slovenia is mainly due to poor construction materials, poor insulation, poor maintenance, old, inefficient, and poorly maintained heating systems and domestic appliances, or a combination of all the above (Tkalec and Živčič, 2014).

This situation is due to a number of reasons, such as lack of savings (especially for the low-income groups), lack of capacity to do construction work, lack of policies to stimulate building retrofitting and the high degree of private ownership, and, implicitly, initiative granted to the owners. Slovenia has a comparatively higher property ownership than the European average. While households living in rental housing have been found to be more likely to be at risk of energy poverty, at least in Western European member states, in 2016, about three-quarters of people in Slovenia lived in their own dwellings, which is 6 percentage points higher than the European average. From an income-based perspective, in Slovenia, just over 50 percent of low-income persons live in properties they privately own

(Primc et al., 2018). Unless solid policies are in place, it is difficult to persuade homeowners to engage in costly retrofitting projects, let alone when that involves community action to retrofit multi-family buildings.

Next to the quality of the housing stock and the fuel-switching phenomenon described above, over the past few years the problem of energy poverty has increasingly been linked to air pollution. Winter months are especially problematic due to the intense burning of fuel wood for heating, mostly in old and inefficient furnaces and stoves (InventAir project, 2018a). In 2014, through the effort of a number of NGOs, nine municipalities were labeled as areas of "excessive pollution" and had to develop plans to tackle the situation (Ministrstvo za okolje in prostor, 2019). This triggered awareness-raising activities within civil society, as well as by public authorities. Nongovernmental organizations established through their discourse the link between energy poverty and pollution and started identifying crosscutting intervention methods. For one example, in 2017 Eco Fund (a public entity in the field of environmental protection that secures financial assistance for initiatives in the field) launched a program to tackle air pollution by financing the replacement of heating devices in socially vulnerable households (Eko sklad, 2020a). For another, in 2019 FOCUS launched a public debate on addressing air pollution through energy poverty alleviation measures.

The state of energy poverty policy

Despite the energy poverty situation described above and the actions undertaken to approach it, some of which will be described below, Slovenia does not have an official definition of energy poverty (Tkalec and Živčič, 2014). Even if there is no official common understanding of the concept, "energy poverty" does occur in a number of government policies. For instance, the National Energy Efficiency Action Plan 2008–2016 (Vlada Republike Slovenije, 2008) highlights the need to tackle energy poverty through an energy efficiency scheme for low-income households. The National Energy Efficiency Action Plan 2014–2020 (Ministrstvo za infrastrukturo, 2015) also points out that there is a need to tackle energy poverty and lists a few possible measures. In the European Cohesion Policy Operational Program 2014–2020 (Vlada Republike Slovenije, 2014), 5 million euro have been earmarked to tackle the phenomenon by subsidizing energy efficiency measures in five hundred low-income households. What is more, the upcoming National Energy and Climate Plan of the Republic of Slovenia (Ministrstvo za infrastrukturo, 2019) puts forward several objectives on energy poverty, including proposing an official definition by 2021 and of an action plan by 2023.

Currently, relevant actors are operating on a common understanding of the notion proposed by the Statistical Office of the Republic of Slovenia, which monitors the phenomenon of energy poverty based on a few indicators. According to this office, energy poverty is "a situation in which a household is unable to provide adequately warm housing (and other energy services, such as hot water, lighting, etc.) at an affordable price" (Statistični urad Republike Slovenije, 2019). In practice, however, due to lack of a common legal understanding and common and comprehensive criteria as to who can be acknowledged as energy poor, only financial indicators are used to identify energy poor households. The immediate effect is that, most often, only recipients of social support qualify for programs offered by Eco Fund, Slovenian Environmental Public Fund that also implements projects related to energy poverty (Eko sklad, 2020b).

This shows that action against energy poverty is possible in spite of the absence of an official definition. Yet, a common definition would generate interventions with a better and more coherent outreach. NGOs have promoted this objective along with a cross-sectorial dialogue on the topic for several years (Živčič et al., 2017). The efforts have resulted in a number of accomplishments, including the draft National Climate and Energy Plan (Ministrstvo za infrastrukturo, 2019), which set out the obligation to elaborate an energy poverty definition and action plan over the following years. The pursuits of the nongovernmental sector will be described below, illustrating some of the projects implemented by various organizations, their individual achievements and how they have, eventually, materialized into national policies. Despite the many steps forward, improvements are needed and these will also be pointed out.

Civil society as the front-runner of energy poverty intervention

In some countries, the concept of energy poverty mainly appears on the public agenda due to researchers and academics exploring the topic, or due to governments taking proactive measures to tackle the struggles of vulnerable households. In Slovenia energy poverty became a priority through a combination of efforts of NGOs and state actors.

The first steps to raise awareness about energy poverty were taken by the National Office for Macroeconomic Analysis, which published a first national report on the topic in 2010 (Vendramin, 2010). It warned about the seriousness of the issue, pointing out that in 2008, 20 percent of households with the lowest incomes were spending at least 15 percent of their available income on energy costs, and urged that appropriate measures be taken to address the problem. At the time, the document did not reach political decision-making circles, let alone the media or the public, at least not in a convincing way.

From grassroots actions to national policies: lessons learned

The leadership role taken by NGOs in the area of energy poverty can be attributed to several factors. First, the action of nongovernmental organizations was encouraged from within the political system after the publication of the report, when it became obvious that public institutions were very limited in their understanding of the issue and in generating action to address it. NGOs, on the other hand, had already gathered expertise both from within the country and at the European level. Second, from the outset, projects implemented by NGOs (such as ACHIEVE implemented by FOCUS), delivered measurable results on the impact of project interventions on household budgets and well-being. These results proved to be sufficiently attractive to decision-makers to scale up some of the measures performed in communities to apply to the national level. Given the practical know-how gained, charities assisted national actions making their implementation much easier and better targeted. Third, nongovernmental organizations influenced the political agenda by persuasion and by engaging multiple parties and the general public in consistent dialogue. They organized conferences, linked relevant stakeholders with one another and with state institutions, and informed the public through the media. Consequently, decision-makers were increasingly charged with greater responsibility in making use of the resources offered to design actions and to tackle the challenges as awareness of them emerged.

A number of projects operated by NGOs were remarkable at generating these resources. The first such project initiated in the country was "ACtions in low income Households to Improve Energy Efficiency through Visits and Energy diagnosis" (abbreviated ACHIEVE), implemented between 2011 and 2014. The initiative was financed through the Intelligent Energy Europe facility. The aim of ACHIEVE was to deliver accessible solutions to households to reduce their state of energy vulnerability across Europe (Živčič et al., 2014). Multiple countries across the continent were targeted. In Slovenia, the project developed a training module for energy advisors, equipping them to perform household visits and recommend the best saving solutions for each individual household based on a set of efficiency performance calculations. Throughout the project, 220 households affected by energy poverty were visited across the country. These households were mainly in Ljubljana, but also in other parts of Slovenia. During the visit, households were offered free kits of energy- and water-saving devices as well as tips on how to improve their consumption behavior. These measures resulted on average in savings of 100 EUR and 320 kg CO_2 emissions per household per year (ACHIEVE, 2014).

After its end, ACHIEVE was followed by REACH, the project on Reducing Energy use And Changing Habits, funded through the same facility. REACH capitalized on the efforts of ACHIEVE, as it embraced its methodology and the lessons learned. But unlike its forerunner, it had a more geographically focused approach as similar initiatives were being implemented also in Bulgaria, Macedonia, and Croatia. Throughout this project an additional 410 households were reached, generating average savings of 75 EUR and 165 kg CO_2 emissions per household per year.

These grassroots projects were a milestone in understanding the specificities of energy poverty in Slovenia, as they collected a large amount of data, and tested a variety of solutions for households. Both ACHIEVE and REACH (ACHIEVE, 2014) (Živčič et al., 2017) trained energy advisors and accompanied them when performing household visits. These training modules instructed long-term unemployed persons as advisors in the case of ACHIEVE, while REACH engaged technical vocational students. Throughout the process, a number of challenges were identified. For the long-term unemployed people, the main challenge was to train them and stimulate their work efficiency given the fact that they were particularly low skilled. But as soon as their skills reached a certain level, they delivered the expected results. However, no solutions were found to continue the work of these trained advisors and to remunerate them once the project was over. Nevertheless, the project succeeded in reintegrating a number of long-term unemployed persons into society by improving their experience and skills, and by encouraging them to remain active. Students, on the other hand, were easy to train, but the fact that schools were unable to continue the energy advising programs after the project end remained an important downside. All in all, students gained useful practical experience and it was established that their interaction with energy poor households, and especially with the elderly citizens, was comparatively more effective than that of those trained from the long-term unemployed group.

In both ACHIEVE and REACH, planning visits to households affected by energy poverty in order to make the needed changes possible was the immediate priority. Although it seemed that free energy advising would be an attractive service for households affected by energy poverty, in practice, this proved to be a cumbersome task. The main obstacle was the lack of trust among households that advisors would really provide their services and kits free of charge. Some also feared that the government or the NGOs involved would use the visits for scrutiny and that they would eventually, for whatever arbitrary reason, disqualify them as recipients of social aid. Some would simply not agree to welcome strangers into their homes. In order to overcome these limitations, the programs experimented with several communication tools, such as leaflets, placards, and vouchers, which presented the visits in an easy-to-understand and attractive manner. Yet, the role of authorized social assistants proved to be the most successful option of all. As vulnerable households had already been in contact with social workers, the trust in the relationship between them was already established. When social workers recommended these assistance programs to them, beneficiaries accepted them with little hesitation.

Visits were implemented in two steps. During the first visit, energy auditors reviewed energy and water bills, conducted a set of measurements regarding lightning, cooking, washing, entertainment, and cooling and heating appliances, and discussed the household's habits with regard to energy use. They also assessed the state of the building (heating system, insulation, windows, position of apartment, etc.) and agreed on the procedure of a second visit. The data collected was entered into an Excel-based calculation tool, which enabled advisors to perform an integrated analysis and shape customized advice for each household.

During the second visit, advisors were equipped with toolkits containing devices to be installed in the households in order to improve energy and water consumption, and with user guidelines. The toolkits comprised energy efficient light bulbs, tap aerators, efficient showerheads, draft-proofing, etc., and were valued at anywhere between 25 and 30 EUR each, but were offered to households free of charge. The advice provided was also adapted to every situation, taking into consideration the specific habits and amenities. Apart from these small changes enabled by the toolkits, households were also informed about the structural problems that would need more serious investment, such as the state of the insulation of the building, heating system, etc. As part of the advice on habits, households were also referred to possible facilities and institutions that could support them in improving their situation further on, such as ways to access insulation funds. Households were also instructed on how to read and understand their energy and water bills.

The impact of visits was calculated on the basis of a methodology developed under two projects (ACHIEVE and EC-Linc). A baseline for each household was established during the first visit, whereas the improvements were noted after the second visit based on the devices installed and their technical specifications (e.g., number of replaced light bulbs or draft-proofed windows). However, in order to understand which of the measures and recommendations worked better than others in the longer run, an additional evaluation was performed through a survey conducted on approximately a third of the households in the program. The survey explored whether the households were still using the devices installed, whether they changed some of their daily habits and whether they had been stimulated to make further investments or to take other steps to tackle energy poverty. The survey was conducted by telephone six months after the second visit in order to allow for the changes to settle in. When compared to the baseline, the evaluation was able to establish the improvements.

It was concluded that the average investment of energy saving devices per household was 27.5 EUR, while their savings over a year amounted to an average of 76.4 EUR. Over an average estimated life span of 7–8 years for the installed devices, households could save some 500 EUR individually. In kWh/year, households saved comparatively more on heating (438.5 kWh/year average per household) than on electricity (262.1 kWh/year average per household). They also saved over 11 cubic meters of water and cut 165.2 kg CO_2/year on emissions per year on average. The standard savings per household amounted to 7 percent of their electricity use, 3 percent of heat use, and 10 percent of water use. All the households visited in the project would eventually save over 1.6 GWh over the lifetime of the installed devices. Note that these calculated savings pertain only to the installed devices; changes in habits, which are much more difficult to quantify, were not considered.

Based on these results, the government designed ZERO, which was the first national energy poverty intervention program. ZERO was run by Eco Fund as an overarching initiative to also tackle air pollution. During its lifetime, the project was constantly adapted to suit the needs observed on the ground. Given its comprehensive objectives, it drew on a number of financial sources. The scheme designed for energy companies to achieve yearly energy savings supported the energy advisors component; the costs of the kits given to households for free to replace their old, less-efficient devices were covered by the Fund for Climate Change run under the Ministry for Environment and Spatial Planning (Stegnar et al., 2018). The results of these grassroots projects show that fairly small technical improvements can generate important savings for vulnerable households. The awareness raising that accompanied these changes is also expected to deliver long-term value. Despite these benefits, ZERO failed to eliminate the core problems households are facing: replacement of old inefficient electrical appliances and proper insulation of buildings against humidity, cold, and excessive heat in summer. These are more substantial investments which ZERO was not equipped to perform, even if its potential would be increased manifold if it were to be accompanied by more thorough structural measures, such as replacing windows, replacing heating systems, or insulating walls (Tkalec et al., 2019).

A number of facilities have been designed to answer to these needs. Eco Fund established a 100 percent financing scheme for the insulation of dwellings in multi-apartment buildings and for the replacement of old stoves and wood furnaces. The program for the replacement of old stoves and furnaces initiated in 2016 was initially intended for recipients of regular social support who owned their dwellings. Later it was extended to recipients of health care allowances, and tenants of municipal and state housing were added to the list of beneficiaries. Despite this, the number of applications remained low, so the program was geographically expanded. At first, it was only meant to target municipalities with the worst air quality in Slovenia. But it was eventually expanded to cover the entire territory. Households were entitled to up to 8,000 EUR individually to carry out heating system improvement (Eko sklad, 2020a).

The refurbishment program in multi-apartment buildings was intended for recipients of regular social support. The program was not meant as a social measure, but rather to remove the typical obstacles that prevent multi-apartment building renovations. Vulnerable households that do not have the necessary means for renovation usually block renovation projects for entire buildings. Hence, the need for the program. Despite its objectives, it is attached to the category of energy poverty interventions.

In 2018 the Jozef Stefan Institute evaluated the impact of the three state-implemented projects described above and established that one of their most important shortcomings was their low potential for outreach. If they are not first properly identified, energy poor households are difficult to reach. This problem is reflected in the low numbers of households finally targeted. Though higher in its impact, ZERO underperformed relative to initial goals, reaching fewer than 300 households per year; the stoves replacement program, on the other hand, only reached 18 households in 2016 and 2017; the refurbishment program reached 152 households between 2010 and the end of 2017. It was established that these results were due to fragile and improperly designed intervention tools and bureaucracy. The programs included a narrow target group, lacked sufficient knowledge about the targeted population, and employed ineffective dissemination tools. A review of all of these projects was deemed necessary to boost their potential (Stegnar et al., 2018).

But the realization of these aims does not solely depend on the availability of resources and on well-written projects. The institutional framework and its capacity to implement are important ingredients. Despite having succeeded to mobilize 5 million EUR from the Cohesion Funds to reduce energy poverty in 500 households by performing retrofitting works, the initiative launched in 2017 has not succeeded in reaching the implementation phase at the moment of this writing. For the most part, this was due to the inaction of the Ministry for Infrastructure, which is an important decision-maker for disbursing these funds. Even after several years, the ministry had not yet prepared the necessary activity agenda to do so. The situation spurred the nongovernmental sector to increase its pressure on decision-makers, which eventually removed the deadlock, prepared the project for operation, and set a completion deadline for 2022 (RTV SLO, 2017).

Engaging stakeholders across sectors: lessons learned

The energy poverty interventions described above were accompanied by activities to promote engagement of local actors, such as municipalities, centers for social work, youth clubs, charities, religious organizations, schools, etc. Activating this engagement was vital to establishing communication with and gaining support for work in local communities. These local actors know their communities and have more appropriate tools to intervene on the ground. To help lay common ground, letters of commitment were acquired from key local actors, while, later on, they were involved in regular communications and information meetings. Their contributions to implementation varied. For instance, the centers for social work and youth clubs facilitated household visits, schools trained their students to become energy advisors, and municipalities facilitated communication with other relevant local actors and offered paid internships to high-school students to perform visits.

Furthermore, all the structural approaches identified through grassroots projects were developed and disseminated through stakeholder engagement.

NGOs enabled numerous policy debates and recommendations and engaged key decision-makers in a persistent manner. They persistently lobbied to add energy poverty to the political agenda. This lobbying took the form of individual meetings with decision-makers where possible programs and measures were discussed: national high-level debates that engaged key experts in a variety of fields, such as energy, health, environment, etc.; and meetings between decision-makers and various social actors to provide these decision-makers with better insight into the social dimension of energy poverty. Because in Slovenia energy poverty is generally perceived as a strongly technical topic and not as a social issue, these interactions were necessary to close the gap between the more technically oriented actors (e.g., those in the ministry in charge of energy issues) and the more social dimension of the decision-making apparatus (e.g., those in the ministry in charge of social affairs), or simply between sectors that would otherwise not communicate with each other. It was hoped that this would result in higher chances for an integrated approach to reducing energy poverty. Also, the topic was highlighted during electoral campaigns with a focus on the use of EU funds for energy poverty actions, which proved to be particularly attractive for politicians (Tkalec and Živčič, 2014).

The high-level debates benefited from frequent media coverage and acted as a simultaneous awareness-raising tool for the general public. Apart from the debates, NGOs organized press conferences related to a variety of topics on the public agenda which they likened to energy poverty, to illustrate the complexity of the phenomenon. The journalists in the specialized media were engaged in a persistent communication on the topic, especially when it related to other subjects of interest, such as the situation of low-income households who cannot afford normal heating, the unsuitable living conditions of some families, and existing options to reduce energy costs.

As many aspects of the national energy and social policies are shaped in Brussels for EU member states, the aforementioned Slovenian NGOs not only added the European perspective to the national discussions, but also contributed the other way around. They made sure that the lessons learned on the ground in Slovenia could have an impact on European policy debates and processes. This objective took shape in the form of co-creation and dissemination activities with representatives of the European Commission and Members of the European Parliament, along with policy recommendations. These efforts also materialized in high-level debates at the core of the European Parliament. This was integrated into larger discussion frameworks, such as the specific challenges of the southeastern European region, and with a larger array of actors, including the energy community, to enable energy market actors and authorities to provide a stable environment that better favored more vulnerable energy consumers.

To assess the effectiveness of the advocacy activities described above, the REACH project, for instance, was evaluated by an external expert (Forbici, 2017). It was established that, even within its short time frame, the project managed to have a number of achievements. The most visible was the establishment of the national program of household visits, ZERO, described above. Furthermore, over 20 key stakeholders – from central to local authorities, schools, or energy experts – were engaged in addressing the challenge of energy poverty in the country, in the long term resulting in a dedicated network of actors. Also, the project resulted in a set of highly effective monitoring and policy-making tools (Tkalec and Živčič, 2014). Lastly, the project contributed to the establishment of the 5 million EUR facility from Cohesion Funds to implement energy poverty actions in Slovenia in the near future.

Several key lessons were learned from advocacy activities. Because energy poverty is a cross-sector challenge, it is important to involve actors in as diverse an array of fields as possible. This is not an easy task given the fact that each of these actors tends to be focused on their own field of work, thus the need for a consolidating force. This can be an NGO, a governmental body, or any other formula. But it has become indisputable that a platform for dialogue and engagement involving a variety of stakeholders should exist if energy poverty is to be effectively and comprehensively tackled. Personal engagement with these stakeholders is of high value, as practice has consistently shown that they can be persuaded much more easily about the relevance of an issue when common interests are established through interpersonal dialogue. From a governance perspective, it is particularly effective to address the local decision-making tier, since they have direct contact with the energy poor households and have better knowledge of local circumstances and resources. Once these principles have been applied, the engagement processes have led to more effective policies, where various stakeholders joined forces to advance the energy poverty agenda in a way that reflected the needs evident on the ground.

Currently, as the work of NGOs has been extended to new relevant topics, such as summer energy poverty in the Mediterranean region, gentrification, or the impact of energy poverty on public health (the 2020 EmpowerMed Horizon project), NGOs have capitalized on all of the previous work, including the networks established to advance these new perspectives. These topics have also been explored by other stakeholders creating a momentum for synergy (Primc et al., 2018). The National Statistical Office, for instance, has prepared an analysis on several new indicators to measure energy poverty in Slovenia (on annual or biannual terms) (Statistični urad RS, 2019).

Conclusions

So far, in Slovenia, energy poverty intervention has been successfully cultivated at the level of nongovernmental institutions, where the concern for the phenomenon first developed, as did the first intervention and policy mechanisms through grassroots initiatives. But despite the effort to establish integrated networks and engage stakeholders to develop suitable methodologies, interventions, and follow-up mechanisms, action to reduce the harms and prevalence of energy poverty remains scattered and inconsistent. The need for more sustained action is obvious. Despite its flexibility and expertise, the NGO sector does not have the strategic, long-term, and data-based perspective the Slovenian government could have, nor its national outreach. NGOs typically depend on short-term funding programs; their outreach and interventions need to be weighed against project budgets. Additionally, prospects for continuity and long-term approaches are marked with uncertainty due to the typical NGO funding process and its dependence on grant-makers' priorities, which are not always tuned to the needs on the ground.

Therefore, the government should be well equipped (in terms of staff, financial resources, know-how, and capacity) to accept the transfer of responsibility for addressing the problem of energy poverty alleviation from NGOs to state institutions. Furthermore, political decision-makers need to understand the importance of participatory policy-making and to develop mechanisms and skills to involve a variety of actors in designing, implementing and monitoring policies related to energy poverty. A variety of actors should be not only involved, but also empowered. NGOs and academic research need a consolidation of their position, including through coordinated funding. Yet another important component of policy-making is the effort to combine the national perspective with attention to local circumstances. To this end, local decision-makers need to be consolidated and allowed the flexibility they need when implementing national programs in a variety of situations on the ground.

Furthermore, suitable and consistent policy-making and instruments are based on the availability of data, that is, not only on a common definition of energy poverty, accompanying indicators, and regular and systematic monitoring, but also on the communication and sharing of existing databases between entities. By including a local perspective and sharply tuned intervention instruments, decision-makers can perform constant policy optimization to adapt to observed circumstances and to design instruments that counter negative externalities and blocked pathways. Data quality and availability is a problem that can only be approached at government level, as is systematic integration of energy poverty measures with a wider set of social and other policy instruments. For instance, national retrofitting programs can counter the adverse impacts of inefficient heating technology on the environment and can thus be approached within a larger policy framework (InventAir project, 2018b). At the same time, national sustainability initiatives, such as energy efficiency programs, should be carefully designed in order to prevent cost increases for the energy vulnerable, and to remove or reduce unnecessary barriers, such as high levels of bureaucracy, as much as possible (Robic et al., 2017). The energy poverty topic should be a part of labor affairs and social policies, such as minimum wages and pensions and access to social support, the improvement of employment opportunities, and professional and industrial transition and requalification (InventAir project, 2018b). In this way the major challenges of energy poverty in Slovenia, which are largely structural aspects, can be prioritized without ignoring the less obvious but still important factors.

To conclude, as noted above, energy poverty is not as extensive an issue in Slovenia as it is in some other countries of the region. This is thanks mainly to the gradual, rather than sharp, transition of the energy market, in turn reflected in energy prices, as well as due to comparatively low income inequality in Slovenia. Yet, a certain part of the population is deeply affected by energy poverty, and this is particularly visible in comparisons to statistics for Europe as a whole. Due to the complexity of energy poverty manifestations, interventions to alleviate energy poverty must be comprehensive.

So far, the combined efforts of NGOs and the government in Slovenia have made progress in learning how to tackle energy poverty, but further steps need to be taken. Energy poverty, being a strongly cross-sectorial and interdisciplinary challenge, calls for involvement of different sectors and disciplines in finding and implementing ways to alleviate the problem. Slovenia finds itself in a decisive moment with regard to its energy poverty strategy. It is crucial for a cooperation platform among a variety of actors (social, energy, environment, health, gender, or technical) to be developed and consolidated. This platform should be government led. It should provide space for dialogue and be receptive to implementing reasonable proposals. In such a way, the country's energy poverty strategy and action plan can begin to be developed as proposed by the emerging National Energy and Climate Plan.

References

ACHIEVE. (2014). *ACtions in low income Households to Improve Energy Efficiency through Visits and Energy diagnosis – Final Publishable Report*. ACHIEVE.

Eko sklad. (2019, December 19). *Eko sklad*. https://ekosklad.si/informacije.

Eko sklad. (2020a). Kurilne naprave na lesno biomaso: Subvencija: www.ekosklad.si/prebivalstvo/pridobite-spodbudo/seznam-spodbud/kurilne-naprave-na-lesno-biomaso/kurilne-naprave-na-lesno-biomaso-subvencija-213 (accessed June 19, 2020).

Eko sklad. (2020b). Zmanjševanje energetske revščine: https://ekosklad.si/prebivalstvo/pridobite-spodbudo/zmanjsevanje-energetske-revscine (accessed June 19, 2020).

Forbici, G. (2017). *Evalvacija Focusovega zagovorništva znotraj projekta REACH*.

InventAir project. (2018a). *InventAir: Methodological Framework for Mapping Energy Poverty and Assessing Its Climate Impacts*.

InventAir project. (2018b). *InventAir: Report on the Energy Poverty and Air Quality Status in the Eastern European Countries*.

Maxim, A., Mihai, C., Apostoaie, C-M., Popescu, C., Istrate, C., and Bostan, I. (2016). Implications and measurement of energy poverty across European Union. *Sustainability*, 8(5), 1–20. www.mdpi.com/2071-1050/8/5/483/pdf.

Ministrstvo za infrastrukturo. (2015, May 21). *Akcijski načrt za energetsko učinkovitost 2014–2020*. Retrieved from www.energetika-portal.si/fileadmin/dokumenti/publikacije/an_ure/an_ure_2020_sprejet_maj_2015.pdf.

Ministrstvo za infrastrukturo. (2019, December 19). *Celoviti nationalni energetski in podnebni načrt Slovenije*. Retrieved from www.energetika-portal.si/fileadmin/dokumenti/publikacije/nepn/dokumenti/nepn_osnutek_avg_2019.pdf.

Ministrstvo za okolje in prostor. (2019, December 19). *Izboljšanje kakovosti zraka*. Retrieved from Onesnažena območja: www.mojzrak.si/onesnazena-obmocja/.

Ministry for Infrastructure RS. (2017). *Akcijski načrt za energetsko učinkovitost do leta 2020 (AN URE 2020)*. Pridobljeno 29. September 2019 iz www.energetika-portal.si/fileadmin/dokumenti/publikacije/an_ure/an_ure_2017-2020-jo.pdf.

OECD. (2019, 12 19). *Income Inequality*. Retrieved from OECD data: https://data.oecd.org/inequality/income-inequality.htm.

OpenExp. (2019). *European Energy Poverty Index (EEPI): Assessing Member States' Progress in Alleviating the Domestic and Transport Energy Poverty Nexus*. OpenExp. Retrieved September 26, 2019, from www.openexp.eu/european-energy-poverty-index.

Primc, K., Slabe Erker, R., and Majcen, B. (2018). *Energetska revščina v Sloveniji*. Ljubljana: Inštitut za ekonomska raziskovanja.

Robic, S., Živčič, L., Kisyov, P., Tkalec, T., and Ilievski, Z. (2017). *Energy Poverty in South-East Europe: Challenges and Possible Solutions – Policy Recommendations*. REACH.

RTV SLO (2017). KODA: Energetska revščina. Retrieved from https://4d.rtvslo.si/arhiv/koda/174505378 (accessed June 19, 2020).

Statistični urad Republike Slovenije. (2012). *Cene v Sloveniji*.

Statistični urad Republike Slovenije. (2019, December 23). *SURS*. Retrieved from Metodološko pojasnilo – energetska revščina: www.stat.si/StatWeb/File/DocSysFile/9200.

Statistični urad RS. (2019, July 8). *Statistični urad RS*. Pridobljeno iz Energetska revščina, Slovenija, 2015: Statistični urad RS.

Stegnar, G., Urbančič, A., Petelin Visočnik, B., Cirman, A., Primc, K., Slabe Erker, R., and Majcen, B. (2018). *Podnebno ogledalo 2018: Ukrep v središču – Energetska revščina (Končno poročilo)*. Ljubljana: IJS.

The Right to Energy Coalition. (2019). *Power to the People: Upholding the Right to Clean, Affordable Energy for All in the EU*. Retrieved September 26, 2019, from https://righttoenergy.files.wordpress.com/2019/02/ep-report-18.02.19.pdf.

Tkalec, T., and Živčič, L. (2014). *Report on the National Situation in the Field of Energy Poverty – Slovenia*. REACH.

Tkalec, T., Živčič, L., and Šavc, K. (2019). *Usmeritve za zmanjšanje energetske revščine in izboljšanje kvalitete zraka v Občini Zagorje ob Savi. InventAir*. InventAir.

Vendramin, M. (2010). *Stroški gospodinjstev za rabo energije v stanovanjih v luči vprašanja energetske revščine*. Ljubljana: Urad RS za makroekonomske analize in razvoj.

Vlada Republike Slovenije. (2008, January 31). *Nacionalni akcijski načrt za energetsko učinkovitost 2008–2016*. Pridobljeno iz www.energetika-portal.si/fileadmin/dokumenti/publikacije/an_ure/an_ure1.pdf.

Vlada Republike Slovenije. (2014, December 11). *Operativni program za izvajanje Evropske kohezijske politike 2014–2020*. Retrieved from www.eu-skladi.si/kohezija-do-2013/ostalo/operativni-programi/op-2014-2020-december-konni.

Živčič, L., Moisan, M., and Tkalec, T. (2014). Project ACHIEVE – Reducing energy consumption in fuel-poor households. *Journal of Energy Technology*, *7*(1), 69–90.

Živčič, L., Tkalec, T., and Robic, S. (2016). Energy poverty: practical and structural solutions for South-East Europe. *Sociology and Anthropology*, *4*(9), 789–805.

Živčič, L., Robic, S., Kisyov, P., Tkalec, T., and Ilievski, Z. (2017). *Fighting energy poverty: achievements and lessons of project REACH*. REACH.

Part II

Energy poverty, energy, and poverty

5 Energy poverty in a subsistence-like economy

The case of North Macedonia

Ana Stojilovska

Introduction

"I use a fuelwood stove to heat one room. When I used to work I used electricity for cooking, but now I am a pensioner and I cook using fuelwood in order to economize. I economize on food in order to be warm at home" (interview with a household of a pensioner couple from Skopje who can partly adequately heat their house, 2017). This quote illustrates the context of subsistence, which is defined as a minimum standard of productive living in society (Sharif, 1986). In a similar fashion, the concept of a subsistence economy has been discussed as an economic level just sufficient to sustain one's family in the case of independent home construction pursued by Yugoslav workers in late socialism and in regard to the popularity of exchange at a Bosnian black market in the early 2000s (Archer, 2018; Jašarević, 2007). North Macedonia can be classified as a "subsistence-like economy" due to the legacy of post-socialistic energy poverty amplified by increasing inequalities and widespread material deprivation. Post-socialistic (post-communist) energy poverty has been described as a reduction of comfort (Tirado Herrero and Ürge-Vorsatz, 2012) and as a form of material deprivation (Bouzarovski and Tirado Herrero, 2017a). Watson and Maitre (2014) argued also in the case of Ireland that energy poverty is not distinct from general deprivation, but rather is an aspect of low standards of living.

North Macedonia is a small and resource-poor former socialistic country which declared its independence in 1991 after the dissolution of Yugoslavia. As with other countries in the Western Balkan region, it is stricken by poverty, income inequality, unemployment, and other forms of social exclusion (Dabrowski and Myachenkova, 2018). North Macedonia is land locked with no domestic oil or natural gas resources, with increasing levels of imported electricity (Macedonian Academy of Science and Arts, 2010; Markovska et al., 2009). The country's gross domestic product (GDP) in 2016 amounted to around 10.7 billion USD at current prices (State Statistical Office, 2018a), with a population of 2.1 million (World Bank, 2019). The data shows inequality and modest living standards, with a Gini coefficient of

32.5 percent in 2017 (State Statistical Office, 2018d), and a 2018 ranking of 82 according to the Human Development Index.[1]

Energy poverty, defined as the inability to attain a socially and materially necessitated level of domestic energy services (Bouzarovski and Petrova, 2015), affects a large part of the population in North Macedonia. The size of the energy poverty issue in the country relates to discussion in the literature about the fact that in some countries in the European Union a large share of the population is affected by energy poverty, while in others it is about certain groups (Bouzarovski and Tirado Herrero, 2017b; Dubois and Meier, 2016). Geographically, the country fits the profile of the "periphery" in the core–periphery distribution of energy poverty in Europe, the periphery being more affected by energy poverty than the countries located in the geographical "core" (Bouzarovski and Tirado Herrero, 2017b). Literature on the geography of energy poverty classifies Central and Eastern Europe as one of the regions with a relatively high share of energy poverty due to a combination of poor building quality, income inequalities, lack of proper infrastructure, cold climates, and inadequate housing, energy, and social welfare policies (Bouzarovski, 2014). Previous research on North Macedonia argues that energy poverty affects not only income poor but also middle-class households (Buzar, 2007). Energy poverty in the country is similar to the case of Hungary, arising as a path-dependent phenomenon resulting from a series of socio-technical lock-ins, such as the infrastructural legacies inherited from the past (Bouzarovski et al., 2016).

This chapter builds on existing academic knowledge about energy poverty in the post-socialist context, and draws on secondary data from statistics and official reports, in combination with primary qualitative data comprised of interviews with stakeholders and households. It describes country-specific characteristics of energy poverty and analyzes its drivers, putting a focus on the context and impacts of the policies relevant to energy poverty. The chapter also investigates the consequences of energy poverty, concluding by summarizing the relevance of energy poverty for ongoing and future energy transition processes. Its original contribution lies in the use of primary data to illustrate the nature of energy poverty and secondary data to explain the social, legal, and environmental determinants of energy poverty and energy injustice in North Macedonia. The novel contribution is the focus on heating technologies, the injustices caused by the energy market's structure, and the impact of policies with relevance to energy poverty, in combination creating a circle of degradation-causing life choices of and for households in energy poverty. The chapter argues that the picture of a subsistence-like economy is completed by a combination of widespread material deprivation, low dwelling quality, lack of developed heat infrastructure, lack of affordable and modern energy technologies and fuels, subsidized electricity and fuelwood prices, the monopoly structure of the energy market, lack of proper support for energy poor people, and well-developed coping strategies and individual solutions used by households.

This keeps North Macedonia durably trapped in energy poverty with far-reaching environmental, health, and development implications.

Materially deprived households

Energy poverty in North Macedonia has three main characteristics. First, a high proportion of households live in precarious material conditions and face problems of material deprivation. Second, energy poverty is closely related to housing ownership leading to individual management of heating and housing; the housing stock consists mostly of detached houses with little thermal insulation. Third, the main heating sources are subsidized electricity and fuelwood used with inefficient combustion technologies resulting in high energy consumption.

The low ability of many households to pay their utility bills is also due to high shares of general income poverty and unemployment in the country. The at-risk-of-poverty rate in North Macedonia, defined as 60 percent of median equivalized income, was 22.2 percent in 2017 (State Statistical Office, 2018d). Not surprisingly, unemployed persons have a higher at-risk-of-poverty rate at 38.7 percent (State Statistical Office, 2018d). The 2017 unemployment rate in the country was 22.4 percent (State Statistical Office, 2018c). There is high inequality in salary levels, with around 77.5 percent of employees receiving a net monthly salary below the average monthly level (State Statistical Office, 2015a; State Statistical Office, 2019a). These inequality indicators are reflected in the subjective reporting of households about their living standards. Only around one-third of households can afford a one-week holiday in a year, while little more than half can afford to eat fish or meat every second day (State Statistical Office, 2018b).

Living at a subsistence-like level means that energy poverty forces households to develop various coping strategies aimed at reducing energy needs or other basic needs in order to afford energy bills. For example, on average the heated area in a home is only 45 percent of the total area of the dwelling (State Statistical Office, 2015b), which means that often only one to two rooms are heated. The low affordability also means that households are unable to afford investments to improve heating, building quality, and household appliances. Household refrigerators with one door are on average 16 years old, washing machines on average 11 years old, and most households use the inefficient incandescent lights (State Statistical Office, 2015b). And because of widespread unemployment, some households rely on only one income, which often is a pension (interviews with households in Skopje, 2017).

Hard-to-heat homes

Households in North Macedonia live mostly in large inefficient houses, leading to high energy costs. On average the dwellings are 83 m^2 and most were built by 1991 (State Statistical Office, 2015b). Around 58 percent of households live in detached houses (State Statistical Office, 2015b). These dwellings also have low energy efficiency as only around 18 percent of households have thermal insulation (State Statistical Office, 2015b). This adds to the difficulty in heating these large and old dwellings. Houses without facade insulation make up a very typical "energy landscape" in cities other than the capital. Research in the cities of Bitola and Kichevo reports that around two-thirds of households in both cities have facades without external insulation (Simonovski and Trajkovska, 2015). On top of that, 15.1 percent of households in 2018 reported that their dwelling had a leaking roof, damp walls, floors or foundation, or rot in the window frames or floor; this is not very high in comparison to Europe (Eurostat, 2019c), but still exposes shortcomings in dwelling quality. Inefficient dwellings impose an additional burden on the already difficult situation resulting from low incomes and high unemployment rate.

Housing ownership also plays a key role in the context of energy poverty. Around 95 percent of households are homeowners (State Statistical Office, 2015b). This means that decisions around housing, including heating and investments in the efficiency of the home and of appliances, are largely an individual matter. This leads to complications when it comes to possibilities and ways in which energy efficiency measures could be undertaken to reduce domestic energy costs. Faced with limited financial means, many households decide either to not engage in energy efficiency or to undertake only partial energy efficiency measures, such as replacement of old windows (interviews with households in North Macedonia, 2017). But partial renovation does not result in the energy savings desired due to the size of the measure (interview with a representative of the private company Knauf Skopje Office, 2017). In addition, in order to save funds, households employ unskilled labor to implement these "improvements" (interview with a representative of the private company Knauf Skopje Office, 2017). This also may result in wasted investment without achieving improvements in the dwelling. The use of unskilled labor is similar to the findings that informal builders in Yugoslavia justified home construction on the basis of the need for physical labor and this resonates with discourses of self-managing socialism (Archer, 2018).

Cheap but "difficult" fuelwood

One of the most distinct factors contributing to energy poverty in North Macedonia is the poor range of high-quality, affordable, and modern energy sources and technologies for heating. This contributes to entrapment of households in a choice between "dirty and cheap" fuelwood or "expensive and available" electricity. The resulting lack of choice for consumers has been highlighted as a consideration for energy injustice by Jenkins et al. (2016). Bouzarovski and Simcock (2017) argue that geographical inequities create spatial differences in energy poverty. The most common type of heating in North Macedonia is fuelwood, followed by resistive electric heating, which in combination are used by 90 percent of households in the country (State Statistical Office, 2015b). District heating is not widespread and only available in the capital city (State Statistical Office, 2015b). At the same time, natural gas supplying households is almost non-existent as only 272 households are connected to the local gas grids, all in the cities of Kumanovo and Strumica in 2018 (Energy and Water Services Regulatory Commission, 2019). Along with the discussed individual arrangements for housing, the lack of developed public heat infrastructure reinforces reliance on individualized types of heating, such as fuelwood or electric heating. Electric heating and fuelwood are the two types of heating most related to energy poverty in the country (interviews with households in Skopje, 2017).

Fuelwood has been noted in the literature as indicative of energy poverty and has been associated with energy degradation in the post-socialistic context as it is a less technologically advanced and more polluting fuel that the energy poor households have shifted toward, as shown in Bulgaria and Hungary (Bouzarovski et al., 2012, Bouzarovski et al., 2016). Households choose fuelwood as a strategy to reduce their energy costs in order to cope with increasing energy poverty (Bouzarovski et al., 2016). Not only is using fuelwood as a heating fuel problematic, but also fuelwood is often used in inefficient stoves. Skopje-focused research shows that around 45 percent of the households in that city use fuelwood for heating and among these users around 71 percent use fuelwood stoves (Skopje se zagreva, 2017), which are usually old and inefficient and often can heat only the room in which they are located (interviews with households in Skopje, 2017).

In addition to its availability and affordability for heating, fuelwood is also attractive due to its product quality and cultural relevance. Once households buy fuelwood (usually for the whole heating season), it is in their possession to use any time of the day or week as suits their needs (interviews with households in Skopje, 2017). The cultural relevance of fuelwood means that it is preferred as it not only economizes on heating costs, but also often replaces other energy services usually satisfied by electricity, such as cooking and preparing hot water (interviews with households in Skopje, 2017). Other strengths of fuelwood are its good warmth in the heated room and the lack of costs for heating outside of the heating season (interviews with households in North Macedonia, 2017). This use of fuelwood has echoes in the subsistence level of energy services and the cultural importance of fuelwood in developing countries. Research on Brazil, Chile, and India points out the importance of culture and community-specific needs in the choice of technology and fuel, in this context fuelwood (Ariztia et al., 2019; Coelho et al., 2018; Jagadish and Dwivedi, 2018).

On the other hand, fuelwood is also often disliked for its environmental implications and because it is considered an outdated "dirty" heating technology. A fuelwood stove is not automatic, does not offer a ready option to regulate the temperature and takes time to warm the room (interviews with households in North Macedonia, 2017). Heating with fuelwood requires a physical presence for repeatedly feeding wood into the stove and is a labor-intensive process (interviews with households in North Macedonia, 2017). Households are aware of the environmental implications of using fuelwood, such as cutting trees and air pollution (interviews with households in North Macedonia, 2017). Although there is a wider consensus that heating with fuelwood is cheap, for some households it is considered expensive (even if the cheapest option available), and some find it difficult to pay for a whole season of fuelwood supply at once (interviews with households in North Macedonia, 2017).

Fuelwood's low price can counterbalance households' energy vulnerability, while high demand for fuelwood is linked to illegal logging. Fuelwood in North Macedonia costs around 57 EUR/m^3 (National Forests, 2019a). According to statistical data, average fuelwood consumption in households that are heated only by fuelwood is 3.97 m^3 in 2014 (State Statistical Office, 2015b). The fuelwood price and the quantity used per household combined show that heating costs are very favorable. However, some households may have costs for fuelwood that are much higher than those based on reported statistical averages, such as 500 EUR for the entire heating season as reported by a household which can afford its heating (interview with a household in Skopje, 2017) – presumably because they use a larger than average amount of fuelwood. The low and in fact subsidized price of fuelwood (interview with an independent expert, 2017) encourages its use. In fact, socially vulnerable persons, such as social welfare recipients, disabled persons, orphans, single parents, and others in similar situations can request a special contract for buying fuelwood: they might pay in 8 installments and they are entitled to a 10 percent discount if they pay in cash (National Forests, 2019b). Due to illegal logging in 2014, 25,230 m^3 of wood was harvested in the country (State Statistical Office, 2015c). Petrova (2014) makes the connection that illegal logging practices in North Macedonia are a case of "roll back" neoliberalism where the state keeps its formal control of the forest sector, but in fact has withdrawn from its management to accommodate private interests.

Expensive but user-friendly electricity

The lack of alternatives can push households towards expensive methods of heating. A factor of energy vulnerability is flexibility and its driving force is the inability to move to a form of energy service provision that is appropriate to household needs (Bouzarovski and Petrova, 2015). In urban areas in North Macedonia without district heating and where fuelwood is not available, households have no option but to heat with electricity (Buzar, 2007). This is similar to the situation in peri-urban areas in Greece, where the lack of gas infrastructure forces households to use other alternatives for heating, often fuelwood (Petrova and Prodromidou, 2019). This lock-in situation has affordability implications. Around 88 percent of households in North Macedonia reported being able to pay their utility bills (State Statistical Office, 2018b). North Macedonia is among the European countries with the highest energy poverty based on the indicator of adequately heating the home, as 24.9 percent of households reported in 2018 that they were unable to keep their homes adequately warm, lower only than Bulgaria (33.7 percent) and Lithuania (27.9 percent) (Eurostat, 2019b).

Households in North Macedonia rely heavily on electricity as additional heating to make up for the spatial and technological limitations of the main method of heating, such as fuelwood stoves. A wide variety of electrical heating appliances are used, and most of them serve as a complementary source of heating with the exception of storage heaters, which are more commonly used as a main type of heating (State Statistical Office, 2015b). Storage heaters are the most commonly used electrical type of heating with a 34 percent share of reported electrical heating appliances (State Statistical Office, 2015b). The storage heater is a typical "socialistic" and inefficient heating technology, which is loaded during the night when the price is low, to emit heat during the day. As a result of their technical inefficiency, storage heaters incur high electricity costs. The inverter air-conditioners which are considered as efficient as heat pumps for households represent only 6 percent of electrical appliances used for household heating (State Statistical Office, 2015b).

Electric heat is often considered a "clean" technologically superior alternative to fuelwood stoves but is more expensive for households to use in comparison. Households assess electric heating as adaptable to individual needs, for example serving well if a specific room, such as the children's room, needs to be constantly warm (interviews with households in Skopje, 2017). It is better than fuelwood with regard to its technological capabilities, such as regulation of temperature and automatic turn on/off, and it is fully for the user to decide when the heating season starts and ends (interviews with households in North Macedonia, 2017). However, using electric heating leads to high energy bills in winter (interviews with households in North Macedonia, 2017). Unlike with fuelwood, the heated room is not sufficiently warm; like with fuelwood, usually one room is heated with one electric heater (interviews with households in North Macedonia, 2017). Some households report that another weakness of electric heat is the quality of indoor air, which is dry compared to with fuelwood heating (interviews with households in North Macedonia, 2017).

Monopolized electricity market

A high proportion of households in North Macedonia live in precarious material conditions, often leading to debts and disconnections. North Macedonia tops the list of European countries with arrears on utility bills according to the European Union Statistics on Income and Living Conditions (EU-SILC) survey with 36.9 percent of households facing this issue in 2018, and closely followed only by Greece with 35.6 percent and Bulgaria with 30.1 percent in arrears (Eurostat, 2019a). Due to unpaid electricity bills, 73,727 consumers were disconnected from electricity services in 2018 based on the latest annual report of the Energy Regulatory Commission (Energy and Water Services Regulatory Commission, 2019); that is roughly 3.5 percent of the population in the country. Organizations analyzing consumer and human rights have been vocal about injustices imposed on households as a result of issues with affording energy bills. The Organization for Consumer Protection has highlighted being charged lump-sum bills and payment of arrears to be the most common challenges regarding electricity and heat supply (Organization for Consumer Protection, 2018). The Ombudsman raised the issue of disconnections from electricity without warning even after only one unpaid bill and the subsequent condition set by the utility to charge the debt in full along with a reconnection fee, causing irremediable damage to households (Ombudsman, 2019). The price of electricity for households is still considered a social category (interview with a representative of the single electricity supplier, EVN,[2] 2017) despite the fact that statistics show that in conjunction with electricity sector liberalization and privatization, the electricity price for households for the period 2008–2018 increased by 82 percent (Energy and Water Services Regulatory Commission, 2008; Energy and Water Services Regulatory Commission, 2018).

The high share of households with arrears also has to do with the monopoly structure of the electricity distribution system, which is composed of just one privately owned supplier and distributor of electricity in which consumers have little trust. Thus, consumers still do not have a real opportunity to choose a supplier, as the efforts to liberalize the electricity sector for households are only recent, via the new Energy Law passed in May 2018. According to this new law, every electricity consumer can choose an electricity supplier (Official Gazette, 2018a). A few years earlier, in 2014, the Macedonian government decided to postpone electricity market liberalization for households, a decision that was criticized for protecting the monopoly of the single electricity supplier (Blazevska, 2014). As a result, in 2016 the Ministerial Council of the Energy Community decided that North Macedonia had failed to ensure that all non-household and household customers were eligible for the purchase of electricity from the supplier of their choice (Energy Community, 2019). The biggest electricity market restructure took place in 2005 when the vertically integrated company unbundled into electricity production, transmission, and distribution, with distribution placed in the hands of a private company in 2006 (Taleski, 2009). The liberalization took place in steps, with companies first able to choose a supplier (Energy and Water Services Regulatory Commission, 2015). The distorted liberalization of the electricity market for households was done under Nikola Gruevski's government when the country's governance was referred to by the European Commission as state capture due to the political dependence of institutions (European Commission, 2015; European Commission, 2016). Electricity market liberalization was set back on track soon after the new government took office. The single electricity utility usually starts legal proceedings against consumers in arrears in order to collect the unpaid debts, which could require the help of an enforcement agent (interview with a representative of EVN, 2017) who has the right to claim finances or property from the household in question. As such, the utility is seen as inflicting injustice via an intrusive "punishing" of households for their inability to pay the bills.

Social policies and vulnerable consumers

North Macedonia has a weak social protection system which is unable to keep households out of income poverty or to protect those affected by poverty. The classical measure to support income poor people is social welfare, which amounts to only 40 EUR monthly for households without any kind of income (interview with a representative of the Platform against Poverty, 2017). Problems arise when a household gets one-time income from another source and welfare is cut and there are no other mechanisms to help that household (interview with a representative of the Platform against Poverty, 2017). The Ombudsman's latest report also highlights that the current social protection system does not respond in a timely manner to the needs of citizens at risk, giving the example of a social welfare recipient family in the city of Veles living in an unsafe illegal dwelling where a fire killed three children (Ombudsman, 2019).

Although there is no national definition of energy poverty, the new Energy Law of 2018 has defined vulnerable consumers, and there is an energy poverty subsidy in place. According to the Energy Law, a vulnerable consumer is a household in which a person lives who, due to their social situation and/or health condition, has the right to use the grid and/or supply of electricity, natural gas, or district heat under certain conditions (Official Gazette, 2018a). The energy poverty subsidy amounts to 16 EUR per month for all types of energy carriers, such as electricity, fuelwood, district heat, and even oil and coal, and recipients are required to pay the energy bills first and then get reimbursed (Official Gazette, 2018b). This subsidy is only for social welfare recipients and as such has too narrow a focus (Stojilovska and Zuber, 2013). This measure is also criticized because it does not contribute to solving energy poverty in the long term since it does not encourage energy efficiency measures (Stojilovska and Zuber, 2013).

Policies for urban planning and building refurbishment

The low efficiency of the housing stock can be explained by the lack of consideration for dwelling quality in past construction. The Skopje earthquake of 1963 destroyed a big portion of the capital city leaving around 140,000 people homeless (Ponjavikj, 2010). Thus, much of Skopje's housing stock was built after the 1963 earthquake. The aftermath involved a period of humanitarian help from many countries as well as a fund for the reconstruction of Skopje (Ponjavikj, 2010). Informally, some buildings are called after the countries which helped their reconstruction, such as the "Russian buildings" or "Czech buildings." Many illegal dwelling were built in the past; thus in 2011 a Law on Procedure with Illegally Built Structures was adopted to legalize illegal dwellings built before this law came into effect (Official Gazette, 2011). These illegal dwellings are dwellings without a building permit (Official Gazette, 2011). A recent statistical report shows that in 2018, however, a total of 955 structures were built illegally (State Statistical Office, 2019b).

Regulations stimulating energy efficiency can play a key role in preventing energy poverty in the future; however, the existing largely inefficient building stock continues to be one of the main drivers of persistent energy poverty in the country. Similarly, Karásek and Pojar (2018) concluded that schemes initiated in the Czech Republic support the decrease of energy consumption, but do not address the issue of existing energy poverty, and work only as preventive measures against energy poverty. Before the national level regulation was adopted, the Municipality of Karposh adopted a local Rulebook on Constructing New Dwellings and Reconstructing Old Dwellings, requiring certain energy efficiency standards over the period 2012–2015 when 244 dwellings of at least energy class C were built or reconstructed (interview with a representative of the Municipality of Karposh, 2017). The national level Rulebook on Energy Characteristics of Buildings was adopted only in 2013, stipulating that new buildings have to be at least energy class C (Official Gazette, 2013) and amending the Law on Construction (Official Gazette, 2014).

Regarding building refurbishment projects, there is a mismatch between the types of measures needed and those offered. Retrofitting is not a panacea for energy poverty and energy efficiency policies need to be carefully designed in order not to reinforce structural or social inequalities (Gillard et al., 2017). At the same time, measuring energy poverty only as a technical and economic condition disregards the complexity of the social condition from which it emerges (Baker et al., 2018). In the context of Southeast Europe it is important to focus on measures for tackling energy poverty, but it is equally important how one approaches the design of such measures (Robic et al., n.d.). The Ministry of Economy has co-financed new windows and solar collectors with beneficiaries chosen on a first-come-first-served basis (interview with a representative of Analytica think tank, 2017). The issue with this measure is the poor choice of policy since many households have already changed their windows (interview with a representative of UNDP, 2017), while the solar collector subsidy is available to all natural persons, so even those who can afford solar collectors are subsidized (interview with a representative of the Energy Efficiency Blog, 2017).

The country's existing energy strategies and programs have the potential to address energy poverty, but this potential remains unfulfilled. The Energy Efficiency Strategy, which identifies measures such as improving energy efficiency in social housing and replacement of fuelwood stoves with, for example, energy efficient stoves, solar systems, etc. is not fully implemented (Stojilovska and Zuber, 2013). Municipalities are obliged to develop energy efficiency programs, but by 2015 one-quarter of the municipalities had not adopted such a program (Stojilovska, 2015). Most of the measures in the programs that have been adopted are about public lighting and insulation of public buildings, while almost no measures to improve heating or for the residential sector exist (Stojilovska, 2015). Many municipalities that adopted energy efficiency programs have not implemented them due to lack of financial resources (Stojilovska, 2015).

Policies for heating replacement

There are some policies aimed at improving the efficiency of household heating, but they mostly target the general household population, which again may subsidize those households which can afford these investments. The City of Skopje co-finances pellet stoves for households in the capital using stoves with fuelwood, oil, coal, or heating oil; applicants have to have bought the pellet stove after the announcement of the subsidy (interview with a representative of Info Center for Energy Efficiency of City of Skopje, 2017). Another example is the recent program of the European Bank for Reconstruction and Development (EBRD) aimed at improving energy efficiency and the availability of renewables for households, which is organized as part grant and part loan, and is distributed through commercial banks (interview with a representative of the WeBSEFF program of EBRD, 2017). This measure is for all natural persons, though they have to be solvent in order to be eligible to apply (interview with a representative of WeBSEFF program of EBRD, 2017). These policies reinforce a focus on individual solutions to reducing energy poverty and improving energy efficiency, as households that cannot afford bank loans or co-finance a pellet stove would probably remain unable to apply for these programs, and continue to rely on themselves for any home improvements.

Policy decision-makers do not necessarily understand the links between climate change, environmental degradation, and energy poverty, and tend not to propose policies linking climate change mitigation with energy poverty alleviation measures. On the one hand, climate policies that increase energy prices through carbon pricing can potentially increase energy poverty levels (Ürge-Vorsatz and Tirado Herrero, 2012), and on the other hand a question is raised as to whether energy poverty alleviation policies could increase final energy consumption and associated CO_2 emissions (Chakravarty and Tavoni, 2013). Often, energy policies do not target adequately vulnerable households and bypass the real needs of these households (Guyet et al., 2018). Fuelwood contributes to indoor air pollution (Bouzarovski et al., 2016) and in an urban context, outdoor air pollution (Papada and Kaliampakos, 2016). Cases from Chile and Lithuania show how the link between climate change action and energy poverty measures was neglected. Although in southern Chile air pollution is known to result from very high levels of fuelwood use for heating, closely linked to inequality and poverty, air pollution plans disregard this social dimension (Reyes et al., 2019). In analyzing energy poverty policies, Kyprianou et al. (2019) show that in Lithuania, reducing the value added tax on district heating and hot water for households was intended to encourage its use in order to reduce air pollution in urban areas but not to support low-income households. Policy examples in North Macedonia which show a similar lack of climate and social policy cohesion are the pellet stove co-financing and the EBRG grant-loan projects. The City of Skopje implements pellet stove co-financing with the aim of reducing air pollution (interview with a representative of Info Center for Energy Efficiency of City of Skopje, 2017), while EBRD aims to reduce CO_2 emissions and achieve energy savings (interview with a representative of WeBSEFF program of EBRD, 2017). Neither of these measures try to connect with social policies and the characteristics of energy poor households.

Economic, social, and environmental costs of energy poverty

Affordability concerns prevent households from accessing existing heat infrastructure and impede construction of future projects on efficient central heating. A planned project for district heating from waste biomass in Kichevo has been difficult to realize because it is not bankable given the currently very low levels of heat consumption related to the fact that dwellings are not fully heated (interview with a representative of GIZ, 2017; interview with representative of the Municipality of Karposh, 2017). No one could afford to connect to this new project (interview with a representative of GIZ, 2017). One key reason is that the electricity price for households is cross-subsidized by the electricity price paid by industry, making an investment in heat infrastructure not bankable (interview with a representative of GIZ, 2017). Due to widespread material deprivation, households in North Macedonia are not able to afford access to gas or district heat infrastructure even if it is available, as it will be more expensive than existing methods of heating – cheap fuelwood combined with well-developed coping strategies. This situation has been defined as poverty of infrastructure, which means lack of access to adequate infrastructure and built environment (Robinson et al., 2018). So, energy poverty is blocking the development of heat infrastructure.

Households develop various coping strategies to manage the limited financial affordability of daily life. Households in energy poverty tend to prioritize the payment of their energy bills (interviews with households in North Macedonia, 2017). In order to be able to afford energy, households reduce their comfort or other basic needs, such as the quality or quantity of food intake (interviews with households in North Macedonia, 2017). The reason for these coping strategies is the lack of sufficient funds to satisfy all basic needs (interviews with households in North Macedonia, 2017). A household that lives in a cold dwelling explains their creative strategy to make up for the lack of warmth: "It is cold because we do not spend much electricity because the bills are huge and we cannot afford them, so we heat little and we cover ourselves with blankets" (interview with a household in North Macedonia, 2017). Brunner et al. (2012) contrast household sufficiency strategies with efficiency strategies, the former being the coping strategies described in this chapter. Households' dilemma between eating or heating is also found in the United Kingdom (Beatty et al., 2014, De Haro and Koslowski, 2013). In another example, some Greek households reduce purchases of food and clothing in favor of covering their energy needs (Papada and Kaliampakos, 2016). Additionally, some studies have found that the Roma community in Hungary is forced to make trade-offs between basic needs, such as paying the electricity bill or for their children's education (Tirado Herrero and Ürge-Vorsatz, 2010).

In extreme cases, challenged with affordability issues, vulnerable households resort to unconventional and unsafe materials for heating. Among the evidenced unconventional heating materials in the case of North Macedonia are tires, wooden boxes for fruit, lath, plastic bottles, car oil, lubricant oils, varnished parquet from renovating, construction waste, heated bedsprings via electricity, used oil, duvets, shoes, and especially the soles of shoes (interview with a representative of the Center for Research and Policy Making, 2017; interview with a representative of Habitat, 2017; interview with a representative of UNDP, 2017; interview with a representative of EVN, 2017). Some Roma households use garbage such as shoes, plastic bottles, bags, old furniture, and rags as fuel in Hungary as well (Tirado Herrero and Ürge-Vorsatz, 2010).

Energy poverty also has serious environmental and health implications. The inefficient use of fuelwood affects energy poverty and causes environmental degradation (OECD/IEA, 2008). Similarly, in Greece fireplaces are used because they are cheap but their intensive use causes smog in urban areas (Papada and Kaliampakos, 2016). Air pollution in North Macedonia is among the highest in Europe, resulting in the fifth highest air-pollution-related death rate in Europe with the main culprit being the widespread use of wood for winter heating by households in urban areas (World Bank Group, 2014). Research indicates that air pollution with particulate matter is responsible for over 1,350 deaths in North Macedonia annually (World Bank Group, 2014).

Conclusions and future implications

Materially deprived households living in inefficient houses and using fuelwood are the most typical profile of energy poor households in North Macedonia. These energy poor households are faced with unemployment, inequality, and lack of high-quality affordable modes of heating. At the same time, they might incur debts and often reduce their energy use or cut back on other basic needs. The choice between fuelwood and electricity is a choice between affordable and highly polluting fuel on one side and expensive but user-friendly heating technology on the other. Often, electrical heating complements the technological limitations of fuelwood. Energy poor households try to cope with their situation at the nexus of the monopolistic electricity market structure, the inadequate social protection system, and the inefficient housing stock inherited from the past. The lack of adequate measures to alleviate income poverty and energy poverty, and the lack of accurate target groups of policies, contribute to the status quo of subsistence living and the consequent development of individual solutions by energy poor households. The synergistic potential between climate and energy poverty measures is neither recognized nor explored. North Macedonia is a subsistence-like economy, shown by the combination of widespread material deprivation, low energy efficiency, missing heat infrastructure, lack of affordable and modern energy technologies, entrapment of households in coping strategies, subsidized energy prices and individual solutions, the monopoly structure of the energy market, and the lack of adequate support for households in energy poverty. Energy poverty contributes to negative environmental, health, and infrastructural impacts. This leads to the conclusion that the country might be durably trapped in energy poverty.

The ongoing market and low-carbon transition has the potential to increase the risk of energy poverty (Boardman, 2010, Bouzarovski and Tirado Herrero, 2017b); thus, it is difficult to be optimistic about prospects for alleviating energy poverty in North Macedonia. Research concludes that meeting the ambitious European Union emission reduction targets would require increasing imports of electricity (Dedinec et al., 2016), which could further increase electricity prices in the country. This raises an additional question as to whether the energy transition will be an inclusive process, or whether it will contribute to increasing the gap between people who are energy poor and people who are not, or those benefiting from the European Union energy transition and those who are not. Since households in energy poverty are preoccupied with their coping strategies and do not receive adequate support, they would likely be left out of the new passive house markets and renewable energy communities, risking that the European energy transition becomes a project for the wealthy.[3]

Notes

1 The Gini coefficient measures income distribution inequality. The Human Development Index measures having a long and healthy life, being knowledgeable, and having a decent standard of living (UNDP, n.d.).

2 Recently the company changed its name to Elektrodistribucija DOOEL Skopje.

3 I would like to thank wholeheartedly the editorial team for their clear guidance, stimulating discussions, and constructive feedback. I am very grateful to the entire ENGAGER research community, which has been my professional "energy poverty" family at a crucial time while I write my PhD. Since most of the work in my chapter comes from my PhD, I am proud to be affiliated with the Central European University, which has shown great strength in defending academic freedom, and my supervisor Professor Michael LaBelle, who has been an incomparable support during the studies. I cannot thank enough my friends and family for their continuous support. Last but not least, I am humbled to have had the opportunity to talk to many energy poor households who shared their untold and touching stories with me. I can only hope that this contribution will enable better understanding of their struggles and the context they live in, and in the long-term shape better policies for alleviating energy poverty.

References

Archer, R. 2018. The moral economy of home construction in late socialist Yugoslavia. *History and Anthropology*, 29, 141–162.

Ariztia, T., Fonseca, F. and Bernasconi, O. 2019. Heating ecologies: Resituating stocking and maintenance in domestic heating. *Energy Research and Social Science*, 47, 128–136.

Baker, K. J., Mould, R. and Restrick, S. 2018. Rethink fuel poverty as a complex problem. *Nature Energy*, 3(8), 610–612.

Beatty, T. K. M., Blow, L. and Crossley, T. F. 2014. Is there a "heat-or-eat" trade-off in the UK? *Journal of the Royal Statistical Society: Series A (Statistics in Society)*, 177, 281–294.

Blazevska, K. 2014. Postponed liberalization – new monopolization of the electricity market. Deutsche Welle.

Boardman, B. 2010. *Fixing Fuel Poverty: Challenges and Solutions*. London; Sterling, VA: Earthscan.

Bouzarovski, S. 2014. Energy poverty in the European Union: Landscapes of vulnerability. *Energy and Environment*, 3, 276–289.

Bouzarovski, S. and Petrova, S. 2015. A global perspective on domestic energy deprivation: Overcoming the energy poverty–fuel poverty binary. *Energy Research and Social Science*, 10, 31–40.

Bouzarovski, S. and Simcock, N. 2017. Spatializing energy justice. *Energy Policy*, 107, 640–648.

Bouzarovski, S. and Tirado Herrero, S. 2017a. Geographies of injustice: The socio-spatial determinants of energy poverty in Poland, the Czech Republic and Hungary. *Post-Communist Economies*, 29, 27–50.

Bouzarovski, S. and Tirado Herrero, S. 2017b. The energy divide: Integrating energy transitions, regional inequalities and poverty trends in the European Union. *European Urban and Regional Studies*, 24(1), 69–86.

Bouzarovski, S., Petrova, S. and Sarlamanov, R. 2012. Energy poverty policies in the EU: A critical perspective. *Energy Policy*, 49, 76–82.

Bouzarovski, S., Tirado Herrero, S., Petrova, S. and Ürge-Vorsatz, D. 2016. Unpacking the spaces and politics of energy poverty: Path-dependencies, deprivation and fuel switching in post-communist Hungary. *Local Environment*, 21(9), 1151–1170.

Brunner, K.-M., Spitzer, M. and Christanell, A. 2012. Experiencing fuel poverty: Coping strategies of low-income households in Vienna/Austria. *Energy Policy*, 49, 53–59.

Buzar, S. 2007. The "hidden" geographies of energy poverty in post-socialism: Between institutions and households. *Geoforum*, 38, 224–240.

Chakravarty, S. and Tavoni, M. 2013. Energy poverty alleviation and climate change mitigation: Is there a trade off? *Energy Economics*, 40, S67–S73.

Coelho, S. T., Sanches-Pereira, A., Tudeschini, L. G. and Goldemberg, J. 2018. The energy transition history of fuelwood replacement for liquefied petroleum gas in Brazilian households from 1920 to 2016. *Energy Policy*, 123, 41–52.

Dabrowski, M. and Myachenkova, Y. 2018. The Western Balkans on the road to the European Union. Bruegel.

De Haro, M. T. and Koslowski, A. 2013. Fuel poverty and high-rise living: Using community-based interviewers to investigate tenants' inability to keep warm in their homes. *Journal of Poverty and Social Justice*, 21, 109–121.

Dedinec, A., Taseska-Gjorgievska, V., Markovska, N., Obradovic Grncarovska, T., Duic, N., Pop-Jordanov, J. and Taleski, R. 2016. Towards post-2020 climate change regime: Analyses of various mitigation scenarios and contributions for Macedonia. *Energy*, 94, 124–137.

Dubois, U. and Meier, H. 2016. Energy affordability and energy inequality in Europe: Implications for policymaking. *Energy Research and Social Science*, 18, 21–35.

Energy and Water Services Regulatory Commission. 2008. Decision.

Energy and Water Services Regulatory Commission. 2015. Annual report for 2014.

Energy and Water Services Regulatory Commission. 2018. Decision.

Energy and Water Services Regulatory Commission. 2019. Annual report for 2018.

Energy Community. 2019. Decisions [Online]. www.energy-community.org/legal/decisions.html.

European Commission. 2015. The former Yugoslav Republic of Macedonia Report 2015.

European Commission. 2016. The former Yugoslav Republic of Macedonia 2016 Report.

Eurostat 2019a. Arrears on utility bills – EU-SILC survey.

Eurostat 2019b. Inability to keep home adequately warm – EU-SILC survey.

Eurostat 2019c. Total population living in a dwelling with a leaking roof, damp walls, floors or foundation, or rot in window frames of floor – EU-SILC survey.

Gillard, R., Snell, C. and Bevan, M. 2017. Advancing an energy justice perspective of fuel poverty: Household vulnerability and domestic retrofit policy in the United Kingdom. *Energy Research and Social Science*, 29, 53–61.

Guyet, R., Živčič, L., Stojilovska, A., Smith, M., Horta, A. and Grossmann, K. 2018. Innovation: Introducing path-breaking perspectives to the understanding of energy poverty. *In:* Bouzarovski, S. (ed.) *European Energy Poverty Policy brief no. 1.* University of Manchester.

Jagadish, A. and Dwivedi, P. 2018. In the hearth, on the mind: Cultural consensus on fuelwood and cookstoves in the middle Himalayas of India. *Energy Research and Social Science*, 37, 44–51.

Jašarević, L. 2007. Everyday work: Subsistence economy, social belonging and moralities of exchange at a Bosnian (black) market. *In*: Bougarel, X., Helms, E. and Duijzings, G. (eds.) *The New Bosnian Mosaic Identities, Memories and Moral Claims in a Post-War Society*. Aldershot: Ashgate.

Jenkins, K., McCauley, D., Heffron, R., Stephan, H. and Rehner, R. 2016. Energy justice: A conceptual review. *Energy Research and Social Science*, 11, 174–182.

Karásek, J. and Pojar, J. 2018. Programme to reduce energy poverty in the Czech Republic. *Energy Policy*, 115, 131–137.

Kyprianou, I., Serghides, D. K., Varo, A., Gouveia, J. P., Kopeva, D. and Murauskaite, L. 2019. Energy poverty policies and measures in 5 EU countries: A comparative study. *Energy and Buildings*, 196, 46–60.

Macedonian Academy of Science and Arts. 2010. Energy Strategy in Macedonia until 2030.

Markovska, N., Taseska, V. and Pop-Jordanov, J. 2009. SWOT analyses of the national energy sector for sustainable energy development. *Energy*, 34, 752–756.

National Forests. 2019a. Our forests. 74 ed. www.mkdsumi.com.mk/admin/documents/nacionalni_shumi_vesnik_br.74.pdf.

National Forests. 2019b. Public announcements [Online]. www.mkdsumi.com.mk/javni_povici.php?a=21&page=1.

OECD/IEA 2008. Energy in the Western Balkans: The Path to Reform and Reconstruction. IEA.

Official Gazette 2011. Law on procedure with illegally built structures 23 ed. www.slvesnik.com.mk/Issues/E7BA62FEE7E4284BBFEB48768BF95B55.pdf.

Official Gazette 2013. Rulebook on energy characteristics of buildings. 94 ed. www.slvesnik.com.mk/Issues/13f04c29ba10487fba87b473c0d7626e.pdf.

Official Gazette 2014. Law on amending the Law on construction. 187 ed. www.slvesnik.com.mk/Issues/d571954a268d4bf6ab73c0bbfb0bdad3.pdf.

Official Gazette 2018a. Energy law. 96 ed. www.slvesnik.com.mk/Issues/dca73ae5850449b2b385b69f4acfd884.pdf.

Official Gazette 2018b. Program for Subsidizing Energy Consumption for 2019. 246 ed. www.slvesnik.com.mk/Issues/14235fb4c7b544beab9a115ee8ebc81b.pdf.

Ombudsman. 2019. Annual report on the respect, promotion and protection of human liberties and rights 2018. http://ombudsman.mk/upload/Godisni%20izvestai/GI-2017/GI-2018.pdf.

Organization for Consumer Protection. 2018. Report for Implemented Activities in 2018. https://opm.org.mk/godisni-izvestai/.

Papada, L. and Kaliampakos, D. 2016. Measuring energy poverty in Greece. *Energy Policy*, 94, 157–165.

Petrova, S. 2014. Contesting forest neoliberalization: Recombinant geographies of "illegal" logging in the Balkans. *Geoforum*, 55, 13–21.

Petrova, S. and Prodromidou, A. 2019. Everyday politics of austerity: Infrastructure and vulnerability in times of crisis. *Environment and Planning C: Politics and Space*, 37, 1380–1399.

Ponjavikj, K. M. 2010. Skopje 1963–2020. *Nova Makedonija*.

Reyes, R., Schueftan, A., Ruiz, C. and González, A. D. 2019. Controlling air pollution in a context of high energy poverty levels in southern Chile: Clean air but colder houses? *Energy Policy*, 124, 301–311.

Robic, S., Živčič, L., Kisyov, P., Tkalec, T. and Ilievski, Z. n.d. Energy poverty in South-East Europe: Challenges and possible solutions.

Robinson, C., Yan, D., Bouzarovski, S. and Zhang, Y. 2018. Energy poverty and thermal comfort in northern urban China: A household-scale typology of infrastructural inequalities. *Energy and Buildings*, 177, 363–374.
Sharif, M. 1986. The concept and measurement of subsistence: A survey of the literature. *World Development*, 14, 555–577.
Simonovski, Z. and Trajkovska, E. 2015. Analysis of the situation and need of improving energy efficiency in the municipalities of Bitola and Kichevo: improving the political dialogue for energy use. Commissioned by the Center for Research and Policy Making. www.crpm.org.mk/wp-content/uploads/2017/12/CELA-KNIGA.pdf.
Skopje se zagreva. 2017. Skopje heats, open data [Online]. www.skopjesezagreva.mk/otvoreni-podatoci/.
State Statistical Office 2015a. Employees and net wages – status October 2014.
State Statistical Office 2015b. Energy consumption in households 2014. Skopje.
State Statistical Office 2015c. Forestry 2014.
State Statistical Office 2018a. Gross Domestic Product 2016.
State Statistical Office 2018b. Household consumption 2017.
State Statistical Office 2018c. Labour Force Survey 2017.
State Statistical Office 2018d. Poverty line Laeken poverty indicators in 2017 final data.
State Statistical Office 2019a. Average monthly net wage paid per employee March 2019.
State Statistical Office 2019b. News Release: Illegally built constructions on the territory of the Republic of North Macedonia, 2018.
Stojilovska, A. 2015. Analysis of the local energy efficiency programs in Macedonia – situation, challenges, solutions. Analytica.
Stojilovska, A. and Zuber, S. 2013. Energy poverty in Macedonia. KAS/Analytica.
Taleski, R. 2009. Electricity reform in the Republic of Macedonia. *Utilities Policy*, 17, 88–101.
Tirado Herrero, S. and Ürge-Vorsatz, D. 2010. Fuel Poverty in Hungary: A first assessment. Central European University.
Tirado Herrero, S. and Ürge-Vorsatz, D. 2012. Trapped in the heat: A post-communist type of fuel poverty. *Energy Policy*, 49, 60–68.
UNDP. Human Development Data (1980–2015) [Online]. http://hdr.undp.org/en/data.
Ürge-Vorsatz, D. and Tirado Herrero, S. 2012. Building synergies between climate change mitigation and energy poverty alleviation. *Energy Policy*, 49, 83–90.
Watson, D. and Maitre, B. 2014. Is fuel poverty in Ireland a distinct type of deprivation? *Economic and Social Review*. Dublin: Economic and Social Research Institute. www.tara.tcd.ie/handle/2262/71985.
World Bank. 2019. The World Bank in North Macedonia/Country Context [Online]. www.worldbank.org/en/country/northmacedonia/overview.
World Bank Group. 2014. FYR Macedonia: Green growth country assessment.

6 Energy poverty as heating poverty in Lithuania[1]

Lina Murauskaite

Introduction

Ever since its independence from the Soviet Union in 1990, Lithuania has made remarkable progress in terms of overall economic performance and living standards, becoming one of the fastest growing economies in Europe (OECD, 2018). However, with 29 percent of the population unable to keep their homes adequately warm (EU-SILC, 2017), it is the EU member state that was found to experience the second-highest level of energy poverty, just after Bulgaria, in a 2015 ranking. The European integration process, to which were added rigorous economic measures aimed at improving the country's macroeconomics during the 2009 economic crisis (sharp cuts in public sector wages, cuts in pensions, tax increases, cuts in social assistance and the tightening of procedures for granting it), had serious repercussions for the well-being of Lithuanian citizens. It also resulted in high inequalities across Lithuanian society. Due to substantial emigration, the population decreased from 3.5 million in 2000 to less than 2.8 million in 2019, while for many of those remaining their ability to deal with daily needs and expenses was significantly reduced.

Under these circumstances, energy poverty increasingly emerged as a social hardship over these three decades, while no public institution has yet proved fit enough to handle it properly. There is little public debate and no legal recognition of the energy poverty concept, while existing policies are insufficient to deal with its trials nationwide. Rather, the top-down approach of European legislation seems to be a way out for the Lithuanian authorities, who continue to consider the issue benign while hoping that by dealing with it in a business-as-usual manner, it will find its own remedy. In particular, energy poverty continues to be approached from a general poverty perspective, while the only separation from this conceptual viewpoint and rapprochement to the sphere of energy is its mention in the Law on Electricity, where the so-called "socially vulnerable consumer" becomes a sector-specific concern. In addition, energy needs go beyond electricity consumption to the heating needs of the population, but there is no mention of it in the Law on Heat or other energy-related legislation. Given the profound changes the energy sector has experienced over time and the social repercussions of these changes across Lithuania, this unsystematic approach is counterintuitive and represents a clear situation of inadequate policy-making where proper measures have been delayed while allowing underlying issues to rest.

Nevertheless, due to the latest European regulations, energy poverty was forced onto the public debate agenda in Lithuania and has, therefore, become an issue of interest for various stakeholders on the market. In January 2019 during the meeting of the Seimas Commission for Energy and Sustainable Development (Seimas, 2019), researchers, policy-makers, and other market stakeholders agreed for the first time to discuss energy poverty, the sources of which they identified to be low household income, inefficient multi-family buildings, and heating prices, in a variety of combinations across Lithuania.

Deriving from this resolution, the current chapter will discuss the income, structural, and market situations that define energy poverty in Lithuania, aiming to point out certain particularities that are especially important to this national debate.

Economic poverty in Lithuania – the central energy poverty factor

As in most cases in the post-socialist region, the concept of energy poverty has gravitated around general poverty without much reflection on the differences, similarities, or impacts of existing interventions in terms of energy access, affordability, and comfort. From a welfare point of view Lithuania remains a paradox. Ever since its integration into the European Union in 2004, Lithuania has displayed strong economic growth, decreasing unemployment, growing wages, and very good public finances, all against the backdrop of widening regional disparities, increasing income inequality, poverty, and social exclusion. Despite displaying low levels of macroeconomic welfare indicators such as the nominal gross domestic product (GDP), in European terms the standard of living for Lithuanians has increased at high speed, reaching a real GDP per capita (which is a measure that takes into account inflation) increment of 1.5 just over the last decade (from 8,710 euro in 2009 to 13,310 euro in 2018; Eurostat, 2019h).

But this macroeconomic development did not necessarily reflect improvements in the personal well-being of households. The unemployment rate dropped by more than a factor of two during the last decade (from 13.8 percent in 2009 to 6.2 percent in 2018; Eurostat, 2019j). Yet, in the context of the high levels of emigration mentioned above, the market faced increasingly low human capital, as a large proportion of unemployed people lack the necessary competences to apply for vacancies. As a result, structural unemployment soared. Monthly minimum wage increased by a factor of 1.7 over the last decade (from 231.70 euro in 2009 to 400 euro in

2018; Eurostat, 2019f) but still remained very low in European terms. This motivated the Lithuanian government to increase the minimum wage to 555 euro in 2019. That increase was accompanied by a personal tax reform, according to which the burden of payment of social insurance was moved from the employer to the employee. In real terms, minimum wages only increased by 30 euros of net income monthly, which rendered the process a technical one without major impact in terms of individual welfare.

Under such circumstances, energy poverty remains hidden behind positive macroeconomic indicators. Despite the optimistic economic statistics, many remain behind.

This assessment becomes self-evident when looking more closely at the social impact of these transformations. One significant indicator is the at-risk-of-poverty rate, which has increased in Lithuania ever since it joined the European Union. In Lithuania the at-risk-of-poverty threshold is set at 345 EUR/month, one of the lowest in the EU. Figure 6.1 shows that almost a quarter of the population (Eurostat, 2019a) are in a state of poverty. As the national statistics presented in Figure 6.3 show, these proportions have not decreased over time. Quite the opposite, some categories have been more affected by increases in energy poverty than others, such as low-income families and those living alone, the unemployed, and retired and disabled people (Statistics Lithuania 2019). For those living in the countryside, aside from other specific obstacles to energy well-being that they might face on a daily basis, the financial burden of energy is comparatively higher: 18.8 percent of the urban and 31.3 percent of the rural population had a disposable income below the at-risk-of-poverty threshold in 2016 (Statistics Lithuania, 2018). This shows that poverty in rural areas of Lithuania is nearly double that in urban areas.

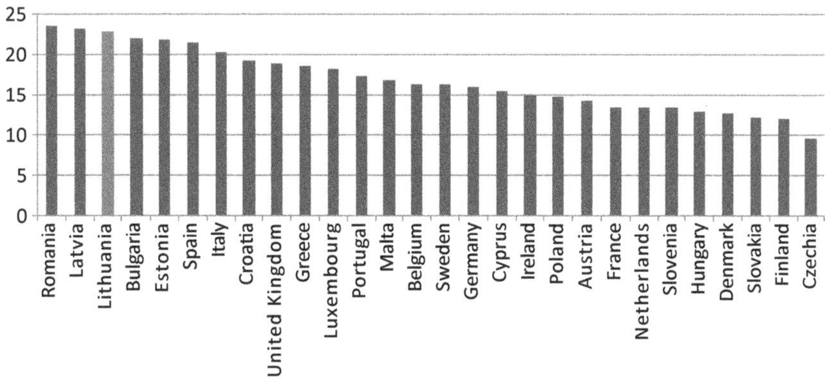

Figure 6.1 At-risk-of-poverty rate for EU countries in 2018.

Source: Compiled by the author based on data from Eurostat (2019a).

The minimum wage provision is far from being a buffer against poverty: after deducting income tax and social security contributions, many individuals are left with a net salary that renders them officially poor. Policies aimed at reducing the tax burden for low-income families are aimed only at minimum income employees and have had a limited impact on their household revenue, while the income tax reform of 2019 left many others with budgets lower than the at-risk-of-poverty rate. People living alone have been particularly affected by these measures. Statistics show that every second single-person household in Lithuania is in a state of poverty (50.7 percent in 2018; Statistics Lithuania, 2019). This group is closely followed by single adults with one or more children (46.8 percent in poverty in 2018; Statistics Lithuania, 2019). Wage differences and fiscal policies are responsible for high income inequality in Lithuania, as demonstrated by the Gini coefficient (Figure 6.2). Moreover, the income level of the most wealthy 20 percent and the poorest 20 percent of people in Lithuania differed by a factor of 7 in 2017 (Eurostat, 2019e). Progressive taxes, capital and property taxation, and stronger redistribution might help tackle inequalities (Lazutka et al., 2019). Although social benefits have increased slightly in recent years, the low progressivity of the tax system limits the impact of redistribution and hence the potential to reduce income inequality and poverty. This accounts for highly disproportionate access to welfare measures and an uneven set of opportunities for well-being.

Compared to other social-demographic categories, pensioners are especially vulnerable to energy poverty from two perspectives: the pension system technically locks senior citizens into a legally secured poverty situation which translates into insufficient means to afford adequate monthly energy services, while their old, long-unrefurbished and energy inefficient

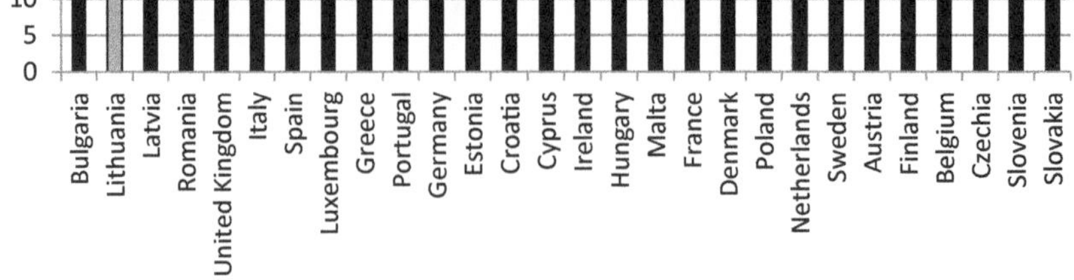

Figure 6.2 EU Gini coefficient of equivalized disposable income in 2018.

Source: Compiled by the author based on data from Eurostat (2019c).

homes generate over-consumption and, therefore, disproportionately high bills. Pensioners represented 38 percent of the at-risk-of-poverty population in 2018 (Figure 6.3); every third Lithuanian pensioner is living in poverty. Despite the increases in minimum and average wages for Lithuania overall, pensions have gone up at a slower pace, standing at 312 EUR (2018) per month on average. Old-age pensions are at a level lower than the minimum wage and even lower than the poverty threshold. Given market dynamics, this led to more than doubling of the percentage of pensioners in an at-risk-of-poverty situation in six years between 2012 and 2018, trapping seniors into a situation of deprivation. From a long-term perspective, the ageing trend of the population against the backdrop of work force emigration makes the current situation unsustainable both in terms of energy poverty and in more general economic terms. It is logical to assume that, in the absence of consistent and well-targeted support programs, reduced financial means make refurbishments difficult for this social category, risking a perpetuation of their situation or even its deterioration in time.

Given the issues described above, there is an important need for welfare support from the state. Assistance programs are designed for people with incomes below the national threshold. However, given the low tax-to-GDP ratio in Lithuania (Eurostat, 2019i), government expenditure on social protection is insufficient. Policies addressing the affordability of heating for low-income households are granted as a compensation for the cost of heating, hot water, and drinking water and are distributed based on the

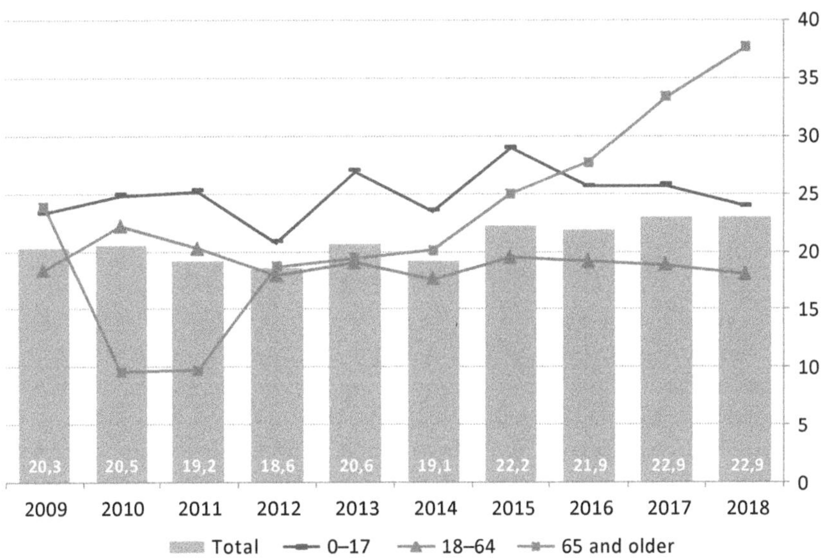

Figure 6.3 At-risk-of-poverty rate in Lithuania.

Source: Compiled by the author based on data from the Statistics Lithuania (2019).

level of household income, the size of households, and the costs of these services. Statistically, over 120,000 persons were part of the compensation scheme in 2018, 58 percent of them being retired or disabled people (Seimas, 2019). The compensation scheme is applied by the local municipalities. Access to these funds still remains limited due to the complexity of the application procedure, which drives some of those who are eligible away from the process (but it is difficult to assess how many).

Additionally, a blanket measure to increase affordability is based on the application of a reduced rate of value added tax (VAT) (9 percent instead of the standard 21 percent) on district heating and hot water. This measure targets effectively all households connected to district heating (i.e., 60 percent of all households), which is 700,000 households in total, located in urban areas. While being important measures for households that could otherwise not afford the cost of heating due to energy inefficient multi-family socialist-era buildings, neither of the two measures manage to target energy consumption itself.

To sum up, three key challenges remain in Lithuania after tax reform: high income inequality, a high poverty rate, and a low rate of social protection. Experts and international organizations (European Commission, International Monetary Fund, and OECD) agree that tax reform has not helped to increase individual purchasing power in the country. In general, decreasing population due to emigration from Lithuania and the increasingly ageing population, deindustrialization, and low incomes determine the specific development of the post-communist country. There are two levels of issues related to energy poverty: urbanization overall and growth in the biggest Lithuanian cities versus the lack of investment and decreasing population in the rural areas. Energy poverty is higher in rural areas, due to a combination of economic and demographic factors. In rural areas, poverty levels are twice as high as in cities. Without a revision of national social policies, further economic growth will not guarantee the improvement of the quality of life of the most vulnerable groups.

Demand side of heating: energy inefficient buildings

Energy inefficient buildings are yet another root of energy poverty identified by the Seimas Commission for Energy and Sustainable Development. The state of the buildings in Lithuania is strongly influenced by its historical heritage. Lithuania declared independence from the Russian Empire in 1918 after the First World War. During the period of interwar independence, the development of industry and transition from an agricultural country to a largely urban one took place. After the Second World War, Lithuania became a part of the Soviet Union. Soviet-era social policy was based on the planned economy, and many public goods, such as housing, heating, and health services, were treated as basic needs and were commonly satisfied needs. Another period of industrialization and urbanization occurred during the 1960s. Unable to meet mass demand for housing, the Soviet Union turned to fast but poor-quality home construction, while a concept of energy efficiency was non-existent (Leonavicius and Genys, 2014). Poor heat insulation was compensated for by unlimited and cheap oil and gas for domestic consumption. Under the market economy this legacy of the planned economy turned out to be one of the most important challenges in terms of energy poverty. Today, 90 percent of multi-family houses, over 80 percent of one and two apartment houses and 92 percent of social houses were built before 1991 (Centre of Registers, 2019) (Table 6.1).

All of these buildings, and blocks of flats, to be very specific, require massive amounts of energy to achieve satisfactory thermal comfort. Multi-family buildings erected between 1961 and 1990 currently house over 60 percent of the Lithuanian population (Streimikiene, 2014), making them a very specific category with respect to energy poverty. Thermal refurbishment of these buildings has been hard to implement and urban energy poverty has remained an issue hard to tackle, as trust between neighbors needed to take common decisions on energy efficiency improvement is scarce.

The Lithuanian District Heating Association (2019) classifies multi-family buildings in four categories based on their specific heat consumption:

1 Seventeen percent of multi-family buildings require a very high consumption of heat. They are generally old buildings, not refurbished, with very poor thermal insulation and old and unbalanced internal heating and hot water systems. The standard heat consumption of these buildings is more than 35 kWh/m^2/month.
2 Fifty-nine percent of the multi-family buildings consume a high amount of heat. They are typically built before 1992, lack insulation, and have had no major refurbishment carried out since they were built. Old and unbalanced internal heating and hot water systems and high hot water consumption have also been reported. The average heat consumption of this category is of about 21 kWh/m^2/month.
3 Seven percent of the multi-family buildings have a medium to low heat consumption. In this category are included new construction with partial refurbishments. Their average heat consumption is if about 15 kWh/m^2/month.
4 Seventeen percent of multi-family buildings are low heat consumers. These are generally new and high-quality buildings or fully refurbished ones. They consume on average 9 kWh/m^2/month.

Based on the classification above, 76 percent of the multi-family building stock is still very poorly insulated and has very high heat energy consumption. Despite efficiency improvements on the district heating production and supply side, households pay high heating bills, precisely due to the high inefficiency of the buildings. As an example, based on 2018 average district

Table 6.1 Housing stock of Lithuania in 2018

Year of construction	*Multi-family houses*		*One and two apartment houses*		*Houses for various types of social groups*	
	units	*percent*	*units*	*percent*	*units*	*percent*
Pre-1940	10,225	26.9	115,264	23.8	286	15.6
1941–1960	3,744	9.8	87,289	18.0	138	7.5
1961–1990	20,285	53.3	203,977	42.1	1,266	68.9
Post-1991	3,825	10.0	78,161	16.1	147	8.0
Total	38,079	100	484,691	100	1,812	100

Source: Centre of Registers, 2019.

heating prices of 5.25 euro cent/kWh on average, owners of a standard 60 m^2 apartment in one of the least energy efficient multi-family buildings pay more than 110 euro monthly for district heat, while owners of a standard 60 m^2 apartment in a very energy efficient multi-family building pay less than 28 euro monthly to meet their energy needs (Lithuanian District Heating Association, 2019).

But interventions on these buildings are far from easy, due to the high private ownership of apartments. Almost 80 percent of all buildings are owned by natural persons and 9 percent of legal persons in Lithuania (Centre of Registers, 2019) (Figure 6.4). Eurostat (2019d) data shows that 90 percent of homes belong to private owners. Many challenges arise from this situation. To be sure, collective action is significantly reduced due to the low degree of trust between neighbors, but also toward public institutions and financial agencies. With such a degree of fragmentation of control, individual capacity is also minimal, especially given how widespread poverty is. Furthermore, as a very specific trait, the Lithuanian housing market is a comparatively static one, generating long-term occupancy of one apartment or house. This is due to a number of issues, including the low trust in the financial market and reluctance to access mortgages. Coupled with low incomes, this reflects in the reduced capacity and incentives for families to invest in energy efficient homes, while choosing instead to spend a higher amount of income on heating. As documented by research performed in the largest Lithuanian cities, there is a great deal of passivity and resistance from the population to invest in renovations due to these dynamics (Lipnevic, 2015). Furthermore, heritage protection requirements, overly long payback periods, a shortage of competent specialists, insufficient state support, etc. are other reasons that explain the difficulty of improving the quality of the housing sector in Lithuania.

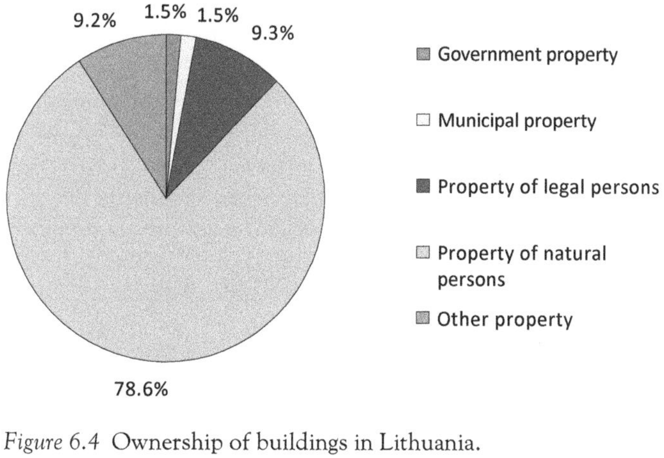

Figure 6.4 Ownership of buildings in Lithuania.

Source: Compiled by the author based on data from the Centre of Registers (2019).

Notwithstanding all these challenges, the government, supported by various stakeholders, has initiated over time several retrofitting programs to multi-family buildings. A first, such initiatives were driven by the World Bank and other donors between 1996 and 2004 as a pilot project to increase the energy efficiency of Soviet-era multi-apartment buildings in Lithuania. Partial refurbishment, without wall insulation, was performed on seven hundred multi-family buildings. Even though the intervention was only partial, heat consumption decreased by a quarter and indoor thermal comfort increased as a result of these actions in all targeted households (Lipnevic, 2015).

The Lithuanian government launched the more substantial "Multi-apartment Building Renovation (Modernisation) Programme" in 2004. The program was more closely targeted toward buildings that were most in need of renovation to increase their energy efficiency, reduce household heating costs and provide low-income families and single residents with the support they needed to overcome the energy burden they were facing. In order to meet these objectives, the program was designed to offer up to 50 percent subsidies on renovations. However, after the financial crisis in 2007 interest rates for bank loans increased significantly and the interest in the program dropped.

A new financing scheme for the Multi-apartment Building Renovation (Modernisation) Programme started in 2010. Funding came through the Joint European Support for Sustainable Investment in City Areas (JESSICA) financial instruments. The European Union allocated 227 million euro to Structural Funds for Lithuania in the form of subsidies of loans for up to 15 percent of investments. The JESSICA-financed scheme secured a fixed 3 percent interest rate for soft loans and up to 100 percent grants for low-income households tin receipt of heating compensation. For all other households, in addition to the 15 percent subsidy coming from JESSICA, a 25 percent subsidy was allocated for investments from the Climate Change Fund, giving a total of 40 percent for investments. Moreover, if most of the multi-family building residents agreed to retrofit their building, low-income households were encouraged to give their consent, otherwise they risked losing compensation for heating costs. Unfortunately, due to low incomes and lack of trust in banks – especially after the financial crisis, and for state authorities in general – program uptake was slow and the total number of renovated multi-family buildings remained unimpressive.

Generally speaking, all of the programs described above had a low impact on energy performance. In terms of efficiency, every modernization would bring the targeted apartment buildings to at least a C energy performance standard according to the Lithuanian regulations, and energy savings to a minimum of 40 percent. But these achievements only reached a minority of buildings. Between 2005 and 2012 only 479 multi-family flats were renovated. The rate increased slightly afterwards, bringing the number of rehabilitated buildings up to 2,369 by the end of 2019 (Housing Advisory Agency, 2019). But the achievements are minimal if compared to the size of the Lithuanian building stock, reaching just 7 percent of its 38 thousand multi-family buildings, most of which are in dire need of intervention (see Table 6.1). This is the point where the structural and economic components of energy poverty in Lithuania collide: incomes are low, while income inequality remains high and households are unable and unwilling to access loans; savings are low and many are incapable of investing in retrofitting out of their own pocket; many social categories are unable to cover their basic needs and increasing thermal comfort is not an attainable option for them; the quality of the residential buildings is low and heat consumption high putting enormous pressure on household budgets and the energy system.

However, the most remarkable achievement is that energy vulnerable consumers were targeted in this Multi-apartment Building Renovation (Modernisation) Programme by providing up to 100 percent grants of investment and interest for energy poor households and tackling energy poverty directly. Moreover, the 3 percent interest rate soft loans and subsidies allocated for investments helped solve the bottlenecks of financing schemes and increased the number of retrofitted multi-family buildings with energy poor households.

Beyond programs that aim at improving district heating, there is additional support for the installation of renewable energy sources during retrofitting of multi-family buildings. This is meant to reduce some of the cost pressures both in household budgets and in the national energy system. The Law on Renewable Energy frames an opportunity to reduce the costs of hot water in apartment buildings by using solar thermal collectors or by switching from district heating to shallow geothermal energy. Today more than 20 multi-family buildings have installed solar collectors and several buildings have adopted shallow geothermal energy production. The impact of these technologies on the final cost of heating is self-evident. However, due to a lack of data, it is difficult to compare real savings between solutions. Also, it is unclear how renewable energy programs have targeted low-income households.

Most of the support schemes, as described, were destined for multi-apartment buildings. However, about 30 percent of the Lithuanian population live in individual houses, most of them located in suburban and rural areas, where poverty is even more widespread. These households rely on biomass and natural gas for heating. Electricity is used mainly for the preparation of hot water and to run home appliances. Heat pumps are used in only a very small number of cases as a comparatively more expensive and less accessible solution. Biomass is the local fuel. It can be secured fairly easily on the local market and its price volatility is slight, being dependent only upon the local market situation. In contrast, all natural gas and about 80 percent of electricity are imported, which makes these commodities all the more expensive, as their prices are influenced by the shaky regional geopolitics. More than 80 percent of individual houses (see Table 6.1) were built before 1991 when energy efficiency regulations were absent. As a consequence, most of these dwellings need substantial investment to reach a reasonable level of efficiency. Household financial resources are generally insufficient to keep houses adequately warm. Retrofitting support schemes for this category of buildings has so far been absent, and thus energy poverty coping strategies have been individual for each household. Recently, the government has set up a financing scheme to target individual households by allocating funds from the sale of carbon emission allowances. It is as yet unclear what the mechanisms would be to target households most in need of support for energy renovations.

Supply side of heating: changes in district heating system for tackling energy poverty

The district heating system is an essential part of the Lithuanian national energy system, and one upon which households and some segments of the economy have depended for 80 years. Over the years the system has endured major changes, but its importance has not altered: 57 percent of heat consumption is secured through district heating, which makes Lithuania one of the countries most dependent on district heating in the EU (Figure 6.5). In urban areas 75 percent of the population is connected to district heating. One fundamental trait of a typical district heating situation for households is that the consumer has practically no control over the heat levels or over

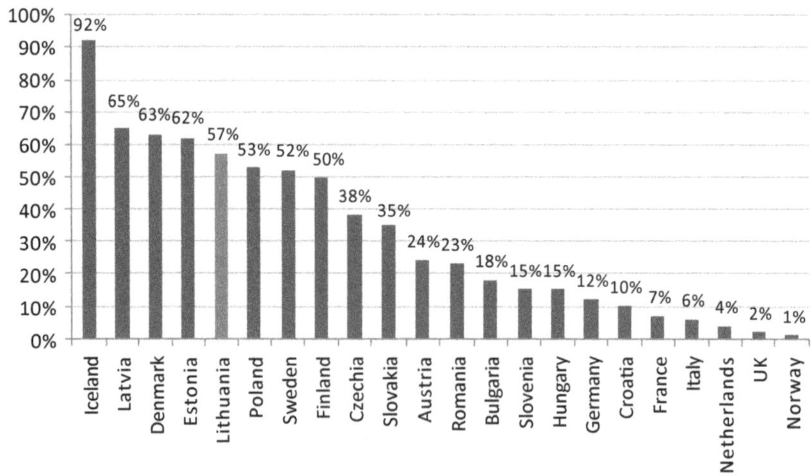

Figure 6.5 Supply of district heating to residents in Europe (percent of residential heat consumption).

Source: Compiled by the author based on data from Euroheat & Power (2015).

heating bills. More specifically, the situation of the vulnerable consumer should be considered within this context.

The construction of the district heating system in Lithuania started in 1939 when the largest medical campus in the country was finished. The system's aim was to secure thermal energy and hot water for the clinics. Moreover, steam was used for laundry, disinfection chambers, kitchens, and sterilizers on campus. After the Second World War and deep into the communist era, the system went through successive development stages. The first cogeneration idea was implemented in Kaunas in 1948 and within just one year the system started to supply hot water for residential buildings there. District heating systems were subsequently introduced in the five largest cities before the end of 1955. After this, the system underwent constant and rapid development, especially between 1960 and 1990 when Lithuania was undergoing accelerated industrialization, urbanization, and increasing construction of the communist style multi-family buildings that remain today, in order to meet the inflow of population that was expected to take up the newly created jobs in the industries. The market model for district heating was based on the centralized planned economy. Government was subsidizing district heating; heat accounting and energy efficiency were practically inexistent. As a result of subsidies, households were not aware of the real costs of heat generation, whereas thermal comfort was mainly acquired through consumption.

But the most important change took place when Lithuania transitioned from a planned economy to a market economy. Independence was regained in 1990, and all district heating companies were integrated into the Lithuanian energy system. Much of the industry collapsed; companies went bankrupt; a wave of uncontrolled disconnections took place across the economy. As a result, the market for district heating plummeted. The process especially hurt industrial cities. Over the next 15 years, consumption of district heating dropped by a factor of 2. Eventually, households started to pay the real price of heat generation and economic stabilization, and energy poverty became more and more evident.

After gaining its independence Lithuania stood on shifting sands. A deep recession followed sudden political changes, economic instability that characterized the country before the introduction of the Lithuanian currency "litas" in 1993, complex and unpredictable changes of the Lithuanian economy, a sharp rise in the cost of primary energy sources, and the economic blockade. Against the backdrop of such economic downturn, in 1994 the government issued the first national energy strategy, which established the main directions for the modernization of an energy inefficient district heating network. The most important changes involved the installation of meters in consumers' substations to decrease the heavy levels of heat loss. Furthermore, fuel reserves were foreseen for each heat-generating unit. The strategy also took the first step toward the development of heat accounting for energy consumers. This was a first initiative to optimize energy consumption in Lithuania.

After the 1996 elections in the Lithuanian Parliament (the Seimas), the government enforced reforms to accelerate market liberalization. At this point, centrally owned and regulated district heating companies were moved to ownership by municipalities, a change that was suggested by the World Bank and International Monetary Fund, who were advising on transition. Under this process of disintegration, the most important question that remained for final consumers regarded the price. Before reform had started, the price of district heating had been subsidized by the Lithuanian Energy Company. After decentralization, prices had to reflect real energy costs. In some municipalities, price increases were enacted at a very high speed. Other municipalities were more cautious and more reluctant to increase household bills. As a result, many district heating companies continued to accumulate debts and, thus, investments were postponed. Consequently, a national body to regulate the energy market and prices was established in 1997.

The costs of these transformations for the population varied across the national territory as a result of the fact that decentralization had different scenarios in every municipality. Some small municipalities decided to abandon district heating due to high financial losses. To compensate, consumers were provided support to install individual gas boilers. Others decided to keep district heating but switch to biomass as fuel. A third group agreed to privatize the systems. The Lithuanian heating market attracted renowned international players, such as Dalkia and Fortum Heat. Private lessees and owners made investment decisions to increase the energy efficiency of the generation and supply network. However, as private capital is profit seeking, prices went up and the new circumstances immediately set off conflicts among politicians, the regulatory authority, and the public in many of these municipalities. Poor households were directly affected by the liberalization process, as low, subsidized prices were not an option anymore, whereas there was no substantial increase in real household budgets. The introduction of heat accounting for every multi-family building by the use of meters in every consumer substation showed the real consumption of heating, whereas consumers remained unable to control their heat use. Continuing to today, heating bills are divided between apartments based on their size (in square meters).

The next important period for the modernization of district heating started in 2003 when the Law on Heat was enacted. The document set the economic, administrative, and political basis of what the Lithuanian district heating sector is today. It opened investment opportunities for the modernization of district heating infrastructure both in leased and municipally owned companies. Based on this law, a long-term incentive pricing was introduced and other financial incentives were given to district heating companies to decrease production costs. These measures had an immediate impact on consumer prices. The long-term incentive pricing is based on costs that are "frozen" at a reasonable level over a base price period of three to five years with only adjustments due to inflation, fuel costs, or other similar objective criteria. Over this period, district heating companies could improve their performance and also make a short-term return. Improved performance indicators were included in the base price of the subsequent period. Thus, consumer prices were allowed to go down from the outset, setting off a win–win situation for consumers and producers for the long term. Energy efficiency of produced and distributed heating also increased significantly.

As a result of all of these transformations, municipalities took on a leading role in heating in relation to both producers and consumers. Municipalities now have a right to set the prices for district heating. They coordinate investment plans and may also give financial support to district heating companies. Municipalities are also closer to the consumers, which enables them to understand their situation better. They are also the main actors in any discussion on energy poverty as they provide social assistance for low-income households.

The modernization of district heating systems accelerated after Lithuania joined the EU in 2004, as the usage of Structural Funds was enabled. Also, the development of the national energy system was better programmed as European Directives became mandatory for Lithuania. Demand and supply were both set for a clear-cut transformation. As a result, a swift change in energy mix took effect in the heating sector, and a series of important renovation projects were instituted to improve consumption efficiency and thermal comfort and reduce the impact of energy costs on the monthly bills of the consumers. The main directives that had a significant impact on the sector are shown in Figure 6.6 and their impact will be discussed below. The European *acquis* was reflected not only in the alignment of both the supply and demand sides with European efficiency priorities, but also in the recognition of the impact of these measures for vulnerable consumers. Thus, the Law on Electricity recognized and defined the socially vulnerable electricity consumer. As pointed out earlier, this was only a fragment and

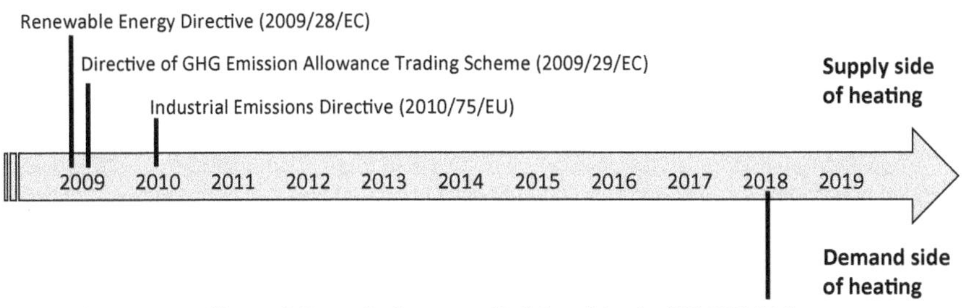

Figure 6.6 Timeline of directives for the supply side/demand side of heating.

Source: Compiled by the author.

rather gave proof of incoherent policy-making, as the same did not occur in the much more important Law on Heat. However, it received much attention in most of the efficiency projects implemented with regards to vulnerable and low-income households.

Combined heat and power plants in Lithuania were traditionally based on natural gas, which was a comparatively expensive fuel on the market, resulting in higher consumer prices. But within a fairly short period of time, five important factors contributed to a revision of the heat-producing energy mix of the country, which in combination had a significant impact on household bills. First, during the financial crisis energy prices soared. Between 2006 and 2009 the price of natural gas multiplied by more than four causing the average price for residential heating to double between 2006 and 2012 (34.7 EUR/MWh excluding VAT in 2006 to 76.6 EUR/MWh excluding VAT in 2012) (see Figure 6.8).

Second, the long-term incentive pricing policy "froze" district heating costs over the base price period to levels that were economically unsustainable for district heating companies. District heating companies running on gas were facing huge losses, which they were unable to entirely reflect in what were already high consumer prices.

Third, the shutdown of the Ignalina nuclear power plant in 2010 also had an impact on the market causing prices to go up (Cogeneration Observatory and Dissemination Europe, 2014). The plant was closed as a result of negotiations with the EU due to its similarities with Chernobyl and in order to prevent another catastrophe. But the decision turned Lithuania into an electricity-importing country. This led to a need for transition as prices increased. At the time, Russia was the main electricity provider for Lithuania. Eventually, as the European internal market developed and Lithuania established interconnections with other neighbors, the price of imported electricity decreased so much that any nuclear investment project became uncompetitive.

Fourth, as specified in the Renewable Energy Directive (2009/28/EC), Lithuania had to reach the mandatory country target of 23 percent renewable energy in final energy consumption by 2020. Under the circumstances it identified biomass as the main source of renewable energy. The most important increase of biomass was established in the district heating sector (Klevas et al., 2013). Large subsidies were allocated to enable the transition. Quotas for other solutions were comparatively reduced.

Fifth, the Industrial Emissions Directive (2010/75/EU) had to be transferred into Lithuanian law. Under these circumstances, two of the biggest combined heat and power plants on natural gas, based in Vilnius and Kaunas, become unfit to compete on the heat market and decided to switch to biomass and waste incineration. Other suppliers soon followed suit. This transition had an immediate impact on final consumer prices, which declined significantly, reaching an average of 49 EUR/MWh (VAT excluded) in 2018. While more than 70 percent of district heating had been generated from natural gas before 2011, after 6 years of transition almost 70 percent of district heating was produced from biomass (Figure 6.7). In parallel to the fuel transition, based on the 2012 National Independence Strategy, Lithuania also decided to build a liquefied natural gas terminal in Klaipeda, which became operational at the end of 2014. Despite high infrastructure costs, the price for final gas consumption decreased by more than 30 percent between 2012 and 2018 (Eurostat, 2019g). Based on these arguments, Lithuania decided to embark upon a fuel transition process that generated important cost reductions for the final consumer.

Besides this fuel transition, a further contextual element that contributed to lowering consumer heat prices is the introduction of competition and market rules on the heat market. Changes in the Law on Heat created this transformation on the market as it allowed independent producers to enter the district heating market and play by partly regulated market rules. In Lithuania district heating plants compete on a monthly basis for the cost of generated heat, which is a unique model in Europe. Priority is given to the lowest price and to renewable energy. Figure 6.8 shows that heating prices decreased from 2012 to 2018, so energy vulnerable consumers have to spend less on heating. What is more, Lithuania became the only country to implement the biomass exchange market (BALTPOOL). District heating companies have been obliged to purchase biomass on the exchange market since 2016, thereby reducing the price by pressure in this market. The average price for biomass increased by almost 30 percent in 2018 due to the

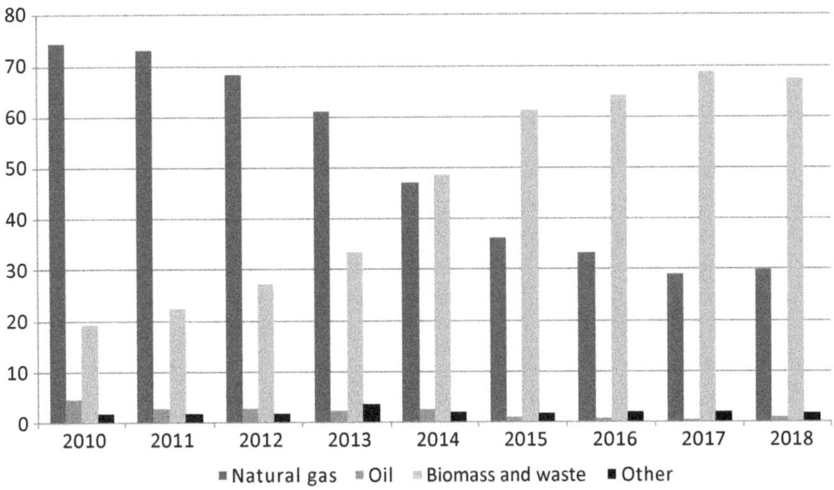

Figure 6.7 Changes in primary fuel used in Lithuanian district heating sector 2010–2018 (percent contribution).

Source: Compiled by the author based on data from the Lithuanian District Heating Association (2019).

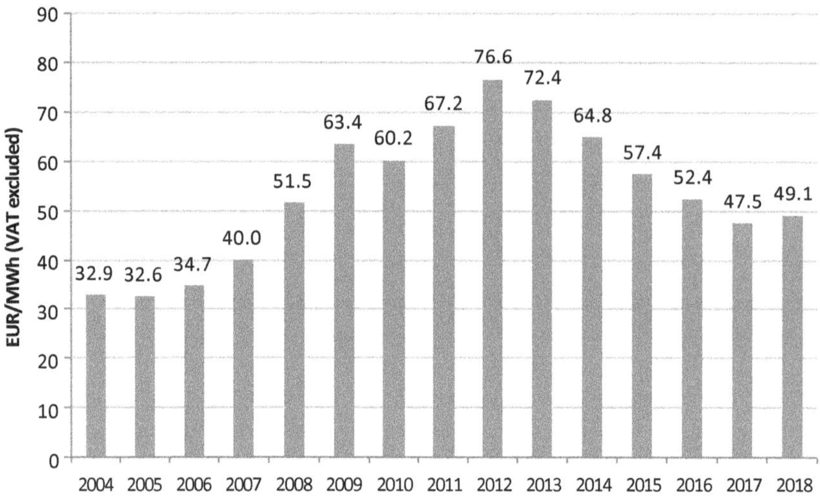

Figure 6.8 Average price of district heating in Lithuania.

Source: Compiled by the author based on data from the Lithuanian District Heating Association (2019).

increase in biomass demand in the international market (National Energy Regulatory Council, 2019) but decreased in 2019. In general, the drive to decrease the district heating price was successful and had a direct benefit in tackling energy poverty in Lithuania.

Despite the change on the supply side, one important challenge remains: with a high number of energy inefficient buildings and corresponding high consumption needs, the total expenditure for heating remains high.

The directive regarding the Greenhouse Gas Emission Allowance Trading Scheme (2009/29/EC) took a step further not only to support the system transition to biomass, but also to enact efficiency and emissions interventions on the demand side. Proceeds from the auctioning of emission allowances were used for the Climate Change Programme in Lithuania, which funded retrofitting projects of multi-family and single-family buildings as well as installation of renewable energy systems for residential use. Based on the government's decision, among the households selected for retrofitting, a number of energy poor dwellings are targeted with priority every year in order to increase consumption efficiency and lower the impact on the budgets of low-income families.

The demand-side-oriented Directive on Energy Efficiency (2012/27/EU) and Directive on Energy Performance of Buildings (2010/31/EU, replaced by Directive (EU) 2018/844) were the main EU pieces of legislation aimed at advancing energy efficiency. Even though in Lithuania discussions around these bills did not refer to energy poor consumers in particular, consultations between the Ministry of Energy and suppliers focused on improving consumption patterns, a measure that would generate important benefits for low-income households. Initiatives were issued to educate households to adapt their energy consumption habits, reduce their monthly energy expenses, and increase indoor thermal comfort. Agreements concerning energy consumers' education and consultation were made between the Ministry of Energy and energy suppliers in order to educate and advise end-users about energy saving measures. To this end, the government and suppliers made a commitment to reduce a minimum of 1 percent of the energy supplied to final customers per annum. Similar agreements were made by heat, electricity, and gas suppliers. This was the first consumer education and consultation model to be implemented across the sector in Lithuania. However, by the end of 2018, only 44 percent of energy suppliers reached the energy savings target (Lithuanian Energy Agency, 2019) as a large part of the agreements took a long time to reach a final stage, while the implementation process is still struggling. Therefore, energy saving levels are below expectations, while the anticipated consumer benefits are still low.

In sum, the Lithuanian district heating sector went through major technical and structural transformations after the collapse of the Soviet Union. Lithuania had to transition from a planned economy to a market economy, from imported natural gas to local biomass, and to comprehensive competition on the supply side of the heat market. All these transformations led to a significant decrease in the price for heat, helping meet a very important need in the population – that of a lower price for thermal comfort.

Conclusions

As a country in the post-communist part of Europe, over the past 30 years Lithuania has focused on enacting a combination of policies in order to increase its energy independence from Russia. This grand mission resulted in a comprehensive approach in terms of the energy market: in technological change, a switch of energy mix, profound efficiency improvements on the supply side of the energy chain, and in persistent thermal enveloping of buildings to improve energy efficiency in household consumption. Though energy poverty is widespread, none of these measures specifically targeted it. It only occasionally prioritized the households most in need of assistance to better afford adequate levels of energy services, while mainly aiming to optimize general energy performance in households whether in energy poverty or not. Energy poverty was only mildly considered in energy legislation, while its true challenges were specifically recognized only in 2019 via the dedicated Seimas Commission for Energy and Sustainable Development. However, as energy poverty's symptoms run deep through Lithuanian society, the real economic welfare, and energy structures, and as energy poverty is hard to separate from many more general social challenges, these far-reaching transformations that have taken place over time have succeeded in addressing energy poverty from many angles, without necessarily being explicitly directed to it. Eventually – also due to European legislation – politics became aware of energy poverty concern and its widespread character.

Today, despite all the structural changes, major inefficiencies in both energy supply and energy demand remain. It is also important to point out that the process of transposition of the European *acquis* into national legislation and discourse is still slow and burdensome. The buildings renovation process is stalling, incomes and inequalities remain high while emigration grows at a steady pace, and macroeconomic engineering hides the real purchasing capacity of the population behind polished numbers. The dynamics of gaining national energy independence, which brought many advantages in terms of energy security and affordability, may yet go too far, as politicians, in the name of energy autarchy, may allow Lithuania to drift apart from the European regional cooperation structures to which it is a party for the purpose of favoring more expensive local biomass. Furthermore, lack of trust runs deep in society and manifests itself at all levels, threatening to generate various effects: lack of trust in European mechanisms may favor more expensive self-reliance options; mistrust in and between institutions, politicians and market actors, such as banks and energy companies, make national policies hard to implement; and mistrust among homeowners makes efficiency programs hard to carry out. This is why energy poverty is persistent after many years of efforts. What is more, the lack of a systematic and consistent approach across all of these efforts has polarized Lithuania even more, making visible many social and geographical discrepancies. The largest part of these projects has almost entirely targeted the urban population and rural areas have been left behind. In rural Lithuania, the percentage of people in poverty is twice as high as in the cities and the population dynamic is characterized by both an ageing population and emigration. Also, no district heating networks supply rural areas, leaving the rural population exposed to high risks of energy poverty.

To sum up, Lithuania has the challenge to deal with a complex task. Many gains have been made but many challenges remain. Energy poverty has been recently identified as a problem in the political discourse but, as mentioned above, its prevalence remains hidden behind good macroeconomic indicators. Balancing between the many challenges while tackling energy poverty remains a knotty task, which will consume Lithuanian society, politics, and economic minds for a long time.

Note

1 I would like to thank the editors, especially Anca Sinea, for their valuable feedback and reviews of this book.

References

Centre of Registers (2019). "Real estate register of the Republic of Lithuania," www.registrucentras.lt.

Cogeneration Observatory and Dissemination Europe (2014). "Final cogeneration roadmap. Member state: Lithuania," www.code2-project.eu/wp-content/uploads/D5.1-Cogeneration-raodmap-NPMS-LT-2014-12-31.pdf.

Directive (EU) 2015/2193 of the European Parliament and of the Council of 25 November 2015 on the limitation of emissions of certain pollutants into the air from medium combustion plants. *Official Journal of the European Communities*, 28.11.2015, L 313.

Directive 2009/28/EC of the European Parliament and of the Council on the promotion of the use of energy from renewable sources and amending and subsequently repealing Directives 2001/77/EC and 2003/30/EC. *Official Journal of the European Communities*, 5.6.2009, L 140.

Directive 2009/29/EC of the European Parliament and of the Council amending Directive 2003/87/EC so as to improve and extend the greenhouse gas emission allowance trading scheme of the Community. *Official Journal of the European Communities*, 5.6.2009, L 140.

Directive 2010/31/EU of the European Parliament and of the Council on the energy performance of buildings. *Official Journal of the European Communities*, 18.6.2010, L 153.

Directive 2010/75/EU of the European Parliament and of the Council on industrial emissions (integrated pollution prevention and control). *Official Journal of the European Communities*, 17.12.2010, L 334.

Directive 2012/27/EU of the European Parliament and of the Council on energy efficiency, amending Directives 2009/125/EC and 2010/30/EU and repealing Directives 2004/8/EC and 2006/32/EC. *Official Journal of the European Communities*, 14.11.2012, L 315.

EU-SILC (2017). "EU Survey on Income and Living Conditions," http://ec.europa.eu/eurostat/web/microdata/european-union-statistics-on-income-and-living-conditions.

Euroheat and Power (2015). "Country by country, 2015 survey," www.euroheat.org.

Eurostat (2019a). "At-risk-of-poverty rate by poverty threshold, age and sex – EU-SILC and ECHP surveys," https://ec.europa.eu/eurostat/statistics-explained/index.php?title=Income_poverty_statistics#At-risk-of-poverty_rate_and_threshold.

Eurostat (2019b). "Electricity consumption and trade, 2017 (GWh)," https://ec.europa.eu/eurostat/statistics-explained/index.php?title=File:Electricity_consumption_and_trade,_2017_(GWh).png.

Eurostat (2019c). "Gini coefficient of equivalised disposable income," http://appsso.eurostat.ec.europa.eu/nui/show.do?wai=true&dataset=ilc_di12.

Eurostat (2019d). "Housing statistics," https://ec.europa.eu/eurostat/statistics-explained/index.php/Housing_statistics#Type_of_dwelling.

Eurostat (2019e). "Income inequality in EU Member States," https://ec.europa.eu/eurostat/web/products-eurostat-news/-/DDN-20190718-1.

Eurostat (2019f). "Monthly minimum wages – bi-annual data," https://appsso.eurostat.ec.europa.eu/nui/show.do?dataset=earn_mw_cur&lang=en.

Eurostat (2019g). "Natural gas price statistics," https://ec.europa.eu/eurostat/statistics-explained/index.php?title=Natural_gas_price_statistics.

Eurostat (2019h). "Real GDP per capita," https://ec.europa.eu/eurostat/web/products-datasets/-/sdg_08_10.

Eurostat (2019i). "Tax-to-GDP ratio up to 40.3% in EU," https://ec.europa.eu/eurostat/web/products-press-releases/-/2-30102019-AP.

Eurostat (2019j). "Unemployment rate – annual data," https://ec.europa.eu/eurostat/en/web/products-datasets/-/TIPSUN20.

Housing Advisory Agency (2019). "Multifamily Buildings Renovation (Modernisation) Programme of Lithuania," www.betalt.lt.

Klevas, V., Murauskaite, L., Kleviene A. and Perednis E. (2013). "Measures for increasing demand of solar energy," *Renewable and Sustainable Energy Reviews*, vol. 27, pp. 55–64.

Law on Electricity of the Republic of Lithuania. *Official Gazette*, 2000, no. 66–1984.

Law on Heat Sector of the Republic of Lithuania. *Official Gazette*, 2003, no. 51–2254.

Law on Renewable Energy of the Republic of Lithuania. *Official Gazette*, 2011, no. 62–2936.

Lazutka, R., Navicke, J., Poviliunas, A. and Zalimiene, L. (2019). "Lithuania is discussing further tax changes to tackle income inequality," European Social Policy Network Flash Report, July 2019.

Leonavicius, V. and Genys D. (2014). "Renovation of multi-apartment houses: social and economic aspects," *Philosophy. Sociology*, vol. 25, no. 2, pp. 98–108.

Lipnevic, A. (2015). "Modernization of apartment buildings in Lithuania: problems and population's initiative," *Public Policy and Administration*, vol. 14, no. 2, pp. 117–193.

Lithuanian District Heating Association (2019). "Review of Lithuanian district heating sector 2018," Vilnius.

Lithuanian Energy Agency (2019). "Implementation of energy supplier agreements," www.ena.lt.

National Energy Regulatory Council (2019). "Heating price statistics," www.vkekk.lt.

National Network of Poverty Reduction Organizations (2018). "Poverty and social exclusion in Lithuania 2018," www.stopskurdas.lt/wp-content/uploads/2018/10/SKURDAS-IR-SOCIALIN%C4%96-ATSKIRTIS-2018.pdf.

OECD (2018). "Launch of the OECD 2018 Economic Survey of Lithuania," www.oecd.org/economy/launch-of-oecd-2018-economic-survey-of-lithuania-july-2018.htm.

Seimas (2019). "Discussion on causes of energy poverty in the Commission for Energy and Sustainable Development," www.lrs.lt/sip/portal.show?p_r=119&p_k=1&p_t=264231.

Statistics Lithuania (2019). "Indicators of poverty," https://osp.stat.gov.lt/informaciniai-pranesimai?articleId=6475033.

Streimikiene, D. (2014). "Residential energy consumption trends, main drivers and policies in Lithuania," *Renewable and Sustainable Energy Reviews*, vol. 35, pp. 285–293.

7 Energy poverty between energy paradigms in Poland

Jakub Sokołowski, Damian Zelewski, Joanna Stępień, and Piotr Lewandowski

Introduction

The footprint of the communist regime over societies, industries, housing conditions, and political approaches is visible everywhere in Central and Eastern Europe; Poland is no exception. From the communist era, Poland has inherited a legacy that is largely associated with – if not inducing – energy poverty in Poland. The energy sector is mostly based on inefficient and outdated hard-coal power plants. The buildings constructed during the communist period are mostly large-scale blocks of apartments with low energy efficiency. After the democratic transformation in Poland, communal heating systems were no longer economically viable. People started to switch to individual heating, using mostly wood and coal, increasing local emissions and failing to account for external costs of those heating systems. Not only do all these changes accumulate to result in a high spread of energy poverty in Poland – in 2017, almost 10 percent of households (about 1.3 million) were regarded as energy poor[1] – but they also pose significant challenges for the future, since Poland now faces the lengthy and bumpy road of energy transition toward a more sustainable but also more expensive energy system.

The goal of this chapter is to provide an up-to-date image of how energy poverty manifests in Poland and how it is dealt with in public discourse. We divide the chapter into two main sections. The first is an overview of expert debates about energy poverty, as well as a survey of relevant instruments of public policy. The second part focuses on future evolutions and covers the political plans of the government (including in the European context), commenting on the newest legal measures and expert recommendations aimed at reducing energy poverty in Poland, taking into account the challenges raised by the process of (just) transition. Finally, the conclusion summarizes the main shortcomings of existing policies in the Polish context and formulates recommendations that may improve their effectiveness.

Debate and previous policy concerning energy poverty

This section covers publications and research by Polish experts and a review of legal instruments dealing with energy poverty. Although the debate on energy poverty in Poland started about two decades ago, even up to now Poland has lacked dedicated support measures for energy poor households. As we explain, existing instruments approach energy poverty only through policies aimed at raising energy efficiency and improving air quality.

Expert debate status

Energy poverty is a relatively new issue in Polish expert debate and public discourse. The discussion on energy poverty in Poland started with the introduction of the definition of "vulnerable energy consumers" into energy law around the year 2000. Later, Kurowski (2008) investigated, for the first time, households at risk of energy poverty in Poland. In 2009, the Energy Regulatory Office published an article entitled "What is Energy Poverty?" reviewing British research in the field of energy poverty (Figaszewska, 2009). However, the governmental strategic document "Energy Policy of Poland until 2030" of 2009 did not mention energy poverty. In contrast, energy poverty is mentioned more than ten times in the "Energy Policy of Poland until 2040" project published in 2018.

Polish academia and nongovernmental organizations recognized energy poverty around 2013, with reports attempting to define the issue and monitor the knowledge of local authorities on the subject (Stępniak and Tomaszewska-Kula, 2013; Pyka et al., 2014; Węglarz et al., 2014). In a survey based on a sample of local authorities (*gmina* in Polish) representatives, only 9 percent of the respondents were familiar with the concept of "energy poverty."[2] The first field research took place in 2013–2015 (Energy Vulnerability and Urban Transitions in Europe "EVALUATE") in Gdańsk (the sixth biggest city in Poland, located in the north of the country). The project focused on experiences of households facing energy poverty (lack of adequate heating, access to electricity). In 2013, the Institute for Structural Research (IBS) started publishing reports and papers on energy poverty focusing on the correct measuring of energy poverty, its regional characteristics, and recommendations for policy instruments. By 2020, IBS has published 14 papers on energy poverty, and are leading in terms of research and expert knowledge in the field of energy poverty in Poland.

Energy poverty support policy

There are no dedicated support instruments for energy poor households in Poland. The issue is only indirectly targeted by other social support policies. The first public instruments to indirectly target energy poor households were introduced in 2014 to help vulnerable consumers, mainly due to the transposition of European directives 2009/72/EC and 2009/73/EC into domestic law. An energy allowance, a new instrument, was granted to people receiving a housing allowance[3] and was supposed to help them in covering electricity bills. The energy allowance is strictly regulated in terms of the housing area (it is practically limited to small apartments and excludes detached houses) and offers a markedly small amount of money to the beneficiaries (equivalent of 2.50–4.40 EUR per month). It is intended to help in paying electricity bills and not heating costs. The effectiveness of the allowance is limited, since energy poverty in Poland is mainly related to high heating costs. Additionally, considering market prices of electricity in Poland, the allowance is too low to help those suffering from energy poverty (Mędrzycki and Szyrski, 2018). There is another regulation dedicated to vulnerable consumers: free installation of pre-paid electricity meters. The administration assumes that pre-paid meters contribute to more-efficient usage of electric energy (Krzykowski, 2014). Finally, there is a fuel allowance with a very strict income criterion (approximately 120–160 EUR per person in a household).

One of the most important issues in tackling energy poverty in Poland is the lack of instruments dedicated to energy poor households. Some welfare benefits aimed at improving their situation exist, but the support is strictly related to the incomes of households. Only half of energy poor Polish households are simultaneously income poor (Sokołowski et al., 2019), which means that the remaining 50 percent of energy poor households are presumably left with no social support.

Policy in the field of energy efficiency improvement

To decrease the scale of energy poverty in Poland, energy poor households need investment support to help refurbish their inefficient houses and apartments. The inefficiency of housing stock results from the fact that multi-family buildings, common in Poland, were mostly built with low-cost technology (sometimes asbestos) in the 1970s and 1980s during communist rule (Figure 7.1). The buildings are energy inefficient and lack thermal insulation. More than half a million energy poor households live in multi-family and detached houses built between 1946 and 1980. The low energy efficiency of those buildings potentially implies higher energy bills and insufficient heating.

Housing stock characteristics differ across Polish regions, ultimately shaping the regional characteristics of energy poverty. In eastern, less developed regions,[4] energy poverty is related mostly to low incomes and high energy spending. The buildings in the east are generally newer, and therefore there are fewer issues with inadequate heating or poor building characteristics. In contrast, in the western regions, houses are older and in worse condition. Therefore, there are more problems with cold homes in

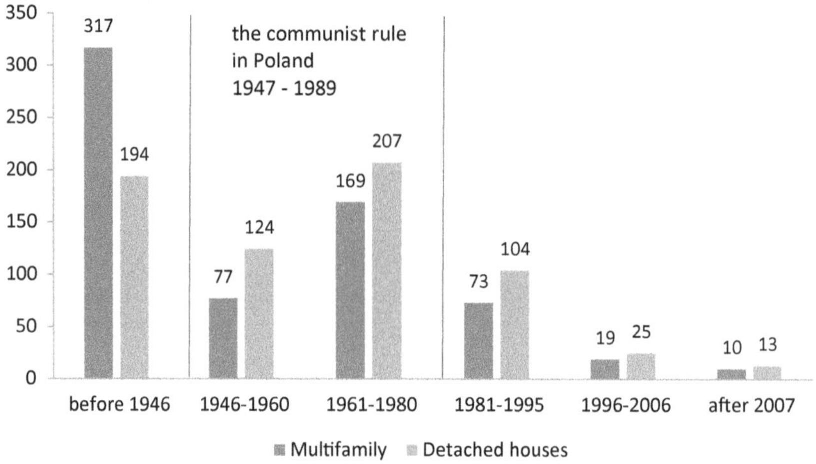

Figure 7.1 Energy poor households in multifamily and detached buildings by the year of building construction (thousands of households) 2017.

Source: Compiled by the authors based on data from the Household Budget Survey (2017).

winter and a wider presence of leaking roofs, damp, or rot. Sokołowski et al. (2019) generated an index which aggregates these various regionally specific dimensions and features of energy poverty and identifies regions and sub-populations in Poland that are in a presumably worse situation than others, as they suffer from more dimensions of energy poverty simultaneously (Figure 7.2).

An additional layer of behavioral issues contributes to energy poverty looming over these households. For example, notoriously, many reduce indoor temperatures by opening windows instead of using the thermostat (Supreme Audit Office, 2019).

Despite the decades over which the challenges raised by poor housing conditions have accumulated, local authorities have not managed to refurbish and improve the state of the building stock on a significant scale (Muzioł-Węcławowicz, 2015), although there have been attempts to implement various solutions, such as loans or thermal retrofitting ventures (Ostańska, 2010).

After the first free election in 1989, Poland changed its economic system from centrally planned to capitalist. During the economic transition period in the 1990s the new government implemented the first programs of replacing benefits for housing associations with loans – or "credits" – for thermal retrofitting. Loans – or "credits" – the symbol of the new, capitalist economy, seemed the more efficient solution. At the end of 1998, a thermal retrofitting bonus was introduced in order to subsidize up to 25 percent of thermal retrofitting loans. The bonus being granted depended on energy

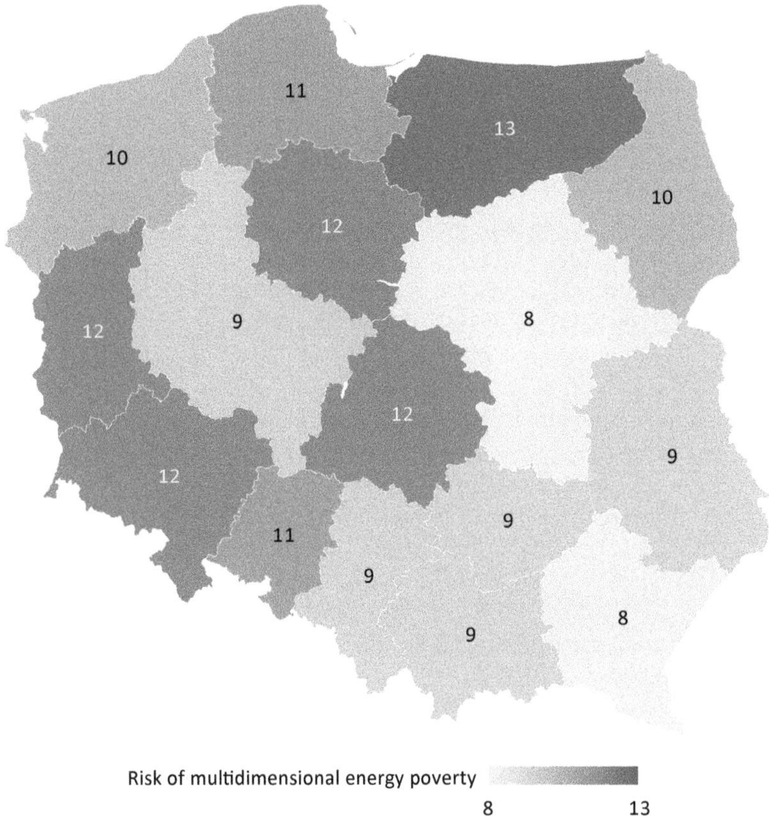

Figure 7.2 Risk of energy poverty by NUTS-2 regions in Poland (percent).
Source: Sokołowski et al. (2019).

audit results. At the beginning, the solution seemed effective, because the number of completed refurbishments grew year by year. From 1999 to 2012, thermal retrofitting bonuses amounted to the equivalent of more than 140 million EUR and were granted for more than ten thousand retrofits (Herbst, 2013). The money from the fund was used mostly by multi-family-building owners (mainly housing cooperatives and associations). Single-family-building owners hardly participated in the program at all, most probably due to the relatively high transaction costs of credits and energy audits (Ministry of Finance and Ministry of Infrastructure and Development, 2015). Finally, energy poor households did not participate in the program because they lacked creditworthiness (Węglarz et al., 2014).

Again, these energy efficiency programs were not aimed at decreasing energy poverty in Poland. The only support instruments that include energy poor households as a target group are instruments aimed at improving air quality, which started being implemented only in 2018, nearly 30 years after the democratic transformation in Poland. We will detail these in the next section.

Policy in the field of air quality improvement

In public debates, the media, and public policy, energy poverty is often seen as a cause of low air quality and high pollution, which are important issues in Poland. Outdated energy sources and cheap fuel may increase environmental strain; nevertheless, the evidence base for this connection is lacking. We can assume this relation is viable because 72 percent of energy poor household use coal as their primary heating fuel, 19 percent use wood, and only 4 percent gas (Lewandowski et al., 2018), so heating is dominated by precisely the fuels most associated with pollution. Additionally, fossil fuels are a fundamental element of the Polish energy system. Greenhouse gas emissions per gross domestic product and carbon intensity in Poland are among the highest in Europe, while Polish air quality is the lowest in the European Union (Dworakowska, 2016). The International Energy Agency (2017) points to household heating systems as a significant source of local air pollution, which in turn raises significant health risks. According to the European Commission and European Environmental Agency, more than forty thousand Poles die due to air pollution every year (Polski Alarm Smogowy, 2016).

Energy poverty has been increasingly acknowledged as an important issue for public policy due to its connection to air pollution. Around 2017, smog became unbearable in the biggest cities of Poland and, as a result, journalists and politicians paid more and more attention to heating sources and fuels related to the quality of dwellings, thus increasingly mentioning energy poverty. That year, 40 articles about energy poverty appeared in popular daily newspapers (ten times more than in 2015 and 2016 combined) (Boguszewski and Herudziński, 2018).

Moreover, the European Commission sued Poland in the European Court of Justice for exceeding air pollution standards, thus forcing the government to take more decisive action. The so-called Anti-smog Act was implemented in 2015. This act enables regional authorities to pass local laws restricting the use of coal in individual households. By mid-2019, 11 Polish voivodships (70 percent) had passed such regulations (Polski Alarm Smogowy, 2019b). Restrictions vary from completely prohibiting the use of solid fuels to restricting the use of specific kinds of fuels or installing stoves which meet specified emissions criteria.

This context stirred debate about energy poor households that could be forced to meet the new regulations, which would result in higher heating costs for those households. Some regions implemented additional support instruments to lower heating costs for the poorest households; for example, this happened in Kraków, one of the most polluted cities in Poland, as well as in Wrocław, both situated in the southern part of the country (Municipal Centre for Social Aid in Cracow, 2019; Official website of the City of Wrocław, 2018).

The Polish government, which had already declared the topic of air quality one of its priorities after it took office in 2015, passed the "Clean Air Programme" in 2018, which we will approach in more detail in the next section. In parallel, the government passed two important regulations regarding the quality of solid fuel stoves and fuels. Some critics argue that, while these regulations may increase air quality, they may also lead to an increase in prices both for more-efficient stoves and for fuel, with a potentially additional negative impact on energy poverty.

The political framing of energy poverty as the reason for poor air quality can be both beneficial and disadvantageous to the broader goal of mitigating the phenomenon. On the one hand, it prioritizes programs of welfare subsidies for energy poor households. On the other hand, energy poverty is sometimes used as an excuse by politicians to hold back the transformation of the energy system in Poland toward a more sustainable one. For example, politicians declare that sustainable (and presumably more expensive) energy will lead to an increase in the scale of energy poverty in Poland (Rolecki, 2019). This approach is also supported by the coal business. The Polish Coal Vendors Economic Chamber published a report in 2019 about the mortality rate caused by energy poverty (Libor, 2019). It became a target of heavy criticism from experts and activists, as an "unreliable, incomplete and misleading" publication, flawed by a spurious methodology (Jędrak and Adamkiewicz, 2019).

Overall, there has been increasing awareness of the issue of energy poverty in Poland, although it still emerges in public debate as a secondary issue associated with other topics, especially air pollution. Additionally, the debate is laced with contradictory interests and a lack of complex analysis of interrelated consequences (e.g., air quality, health, energy system transformation).

Energy poverty after 2020: plans and recommendations

This section deals with policy plans and programs for energy transformation in Poland. These strategic plans usually refer to energy poverty. Additionally, we present expert recommendations for changes in addressing the energy poverty issue in Poland.

Strategic governmental plans

The main strategic document that currently refers to energy poverty is "Energy Policy of Poland until 2040" (which is still a draft as of February 2020). It mentions the reduction of energy poverty through thermal retrofitting of the housing stock as one of the energy sector's priorities. Although policy-makers emphasize the need to limit solid fuel use by individual households, they highlight that the process should be gradual and consider the risk of energy poverty. The document states that

> first, special support in thermal retrofitting programs for the poor households is a key measure against energy poverty. … Second, the households should receive support for changing the heating source in order to stop burning waste and low-quality solid fuels. Currently, the energy allowance supports the vulnerable consumers, but it is not an efficient response to the issue of energy poverty, therefore, we will investigate new effective measures against energy poverty.
>
> (Ministry of Energy, 2018)

Unfortunately, these actions do not have any defined timeframe.

Second, we mention the governmental "Clean Air Programme" (with a special component targeted to energy poor households called "Stop Smog"), implemented for the years 2018–2029 with a budget of approximately 24 billion euro. The program allows subsidies and loans for installing new heating sources and thermal retrofitting of buildings. It has different levels of subsidies related to income thresholds, with the highest subsidy of 90 percent set for households with the lowest incomes. After the introduction of the program, experts noticed some of its practical dysfunctions, such as inefficiency of processing the requests of potential beneficiaries (Polski Alarm Smogowy, 2019a). Consequently, the government introduced the first modifications in 2019 (Ciepiela, 2019). Still, the program may have the countereffect of increasing the costs of heating if a beneficiary decides to install a new heating source without a thermal retrofit, since the prices of clean fuels are generally higher than the prices of coal and wood. This scenario is especially likely in the case of energy poor households, as changing the heating source is cheaper than insulating or rebuilding the house. The "Stop Smog" section generates a mechanism through which local authorities are the beneficiaries of support and they then target locally the households in need of help. Energy poor households are allowed full financing of the retrofit investment and heating source installation. However, until the beginning of 2020 only 8 (of nearly 2,500) local authorities have entered the program.

Moreover, other support instruments, such as thermal retrofitting relief, have been designed with the purpose of achieving the targets set in strategic documents. Some of them were introduced only at the beginning of 2019, so at the time of writing this chapter it is still too early to assess their effectiveness and impact on the scale of energy poverty in Poland. The thermal retrofitting relief gives taxpayers an opportunity to deduct from their income up to an equivalent of 12,000 euro for covering the costs of thermal retrofitting of a single-family house and installing a new heating source (Wielgo, 2019).

Unfortunately, energy poor households are rather excluded from this instrument due to their low incomes and lack of loan opportunities.

Energy poverty and just transition

Energy poverty alleviation measures need to be connected to the just transition of the energy sector in Poland. The Polish energy industry is one of the most emissive in the world because of its high reliance on fossil fuels. In line with the effort of European Union members, Poland must reduce energy-related emissions and transform the energy industry into a more sustainable system based on renewable energy. This transformation will be costly (140 billion euro as estimated by the Polish government), which can lead to further social inequalities and unequal access to clean and efficient energy. Individual households need to invest in energy efficient technologies on their own; this possibly increases the risk of energy poverty, despite the support provided by the EU. Prohibiting combustion of solid fuels without any direct support mechanisms may have a similar effect (Libor and Bouzarovski, 2018). Finally, an ambitious heating sector transition may require intensive thermal retrofitting projects, which without support mechanisms may increase the risk of energy poverty by 10 percent (Forum Energii, 2019).

Some additional dimensions need to be considered under the general framing of a just transition. The first of these dimensions is the link between energy poverty and energy prices. Nominally, energy prices in Poland are among the lowest in the EU. However, adjusted to purchasing power parity, Poland is among the ten most expensive countries in terms of electricity prices (Eurostat, 2018). The burden of paying electricity and heating bills on households in Poland is greater than for the residents of most other European countries. In this situation, any price increase carries a risk of increasing the scale of energy poverty. Changes in the country's fuel mix and the energy transition process may cause considerable price increases affecting individual recipients, regardless of their economic situation.

Second, intensive changes in the fuel mix might result in significant turbulence for the Polish mining sector and related industries. Since WWII, coal has been the only available fuel in Poland that could meet the country's growing energy demand. Relatively small national reserves of hydrocarbon fuels and their high prices on international markets resulted in the dominance of coal technologies. After the geopolitical changes of 1989, a shift toward sustainability and energy efficiency was high on the agenda. Although the transformation of the energy sector is highly prioritized today, changes to the fuel mix should be conducted gradually and with due consideration of the socio-economic context. Despite closing outdated coal power plants and mines and reduction of coal-related employment by about 80 percent compared to 1989, Poland is still transforming the energy sector.

The process of change in the conventional energy sector is primarily related to the need to reduce energy costs and CO_2 emissions. The reduction in employment within the energy sector potentially affects the Silesia region the most. Hard-coal mining employs more than 70,000 people, which represents a significant share of total employment in this sector in the EU and about 86 percent of hard-coal mining in Poland (Polish Electricity Association , 2018). The restructuring programs currently being implemented take into account the conversion of employees leaving the sector and their inclusion into other sectors of the economy. Those changes may increase the risk of energy poverty in coal-intensive regions, as energy prices and the unemployment rate may increase due to the transition process.

Conclusions and recommendations for reducing the scale of energy poverty in Poland

The efficient identification of energy poor households and of those at risk of energy poverty is crucial for a well-designed support system. For this purpose, a definition of energy poverty needs to be introduced in the legislation. As of the writing of this chapter, Polish authorities have not yet decided whether and how to introduce an official, national energy poverty definition in any legal documents, although the first steps have been made. In consultation with the government, the Institute for Structural Research proposed the following definition of energy poverty, aimed at capturing local context specificities and allowing for the creation of knowledge-based policy:

> Energy poverty occurs when a household is unable to afford the energy needed to provide its members with adequate warmth, cooling, lighting, and appliance use due to a combination of factors, which may include having a low income, high energy expenditure requirements, and a home with low levels of energy efficiency.
>
> (Sokołowski et al., 2019: 8)

With respect to the policies currently implemented, we highlighted earlier that the rules of a thermal retrofit support are a barrier for households at risk of energy poverty. Energy poor households need full financing to make the necessary upgrades to their houses; therefore, we believe they should be offered subsidies rather than loans, even if loans have favorable financing rules. The support programs should not be limited only to detached houses, but should include also multi-family and municipal buildings. Finally, local authorities could act as a backing institution for energy poor households in obtaining financing for thermal retrofitting interventions through the dedicated loans (Rutkowski et al., 2018).

Another possibility is financing investments in energy efficiency through nongovernmental organizations. The mechanism could be especially applicable in the poorest regions, where a large number of households are incapable of covering the costs of renovation. Selected nongovernmental organizations could support the implementation of thermal retrofitting projects and installation of energy efficient heating sources. This mechanism, financed from public funds, would reach energy poor households that are unable to go through any administrative procedures related to obtaining a thermal retrofitting bonus (Rutkowski et al., 2018).

Polish policy-makers should reform current social policies to finance clean energy sources and efficient fuel for energy poor households through those policies. Energy poor households could receive "energy cheques," similarly to a solution applied in France. Additionally, energy poor households could receive an increase in their fuel allowance during the heating season (October–March). Finally, Poland needs to reform the energy allowance and simplify the granting procedure to reduce transaction costs. At the moment, the energy allowance is granted only to households receiving a housing allowance. If the instruments are divided into separate support mechanisms, their efficiency will be improved by targeting more households in need of help. For example, the current criteria for the housing allowance eliminates the rural population from the group of potential recipients of the energy allowance. Finally, a useful tool may be energy advisors, who already exist in all 16 regions, but who might be further empowered in order to become more effective. They can train local energy consultants to help energy poor households in performing small-scale renovations or educate local communities on efficient and clean heating techniques (Węglarz et al., 2014; Rutkowski et al., 2018).

Finally, an important dimension of energy poverty that is yet to be addressed in Poland, but that will likely gain awareness as a result of the energy transition, is its impact on health. Living in an inadequately heated house can lead to adverse health-related consequences. Following Thomson et al. (2017), self-reported poor health and poor emotional well-being among energy poor people is unequally distributed across European countries, with the highest overall rates of poor physical and mental health in Central and Eastern Europe. However, unfortunately, a significant gap in research on energy poverty and its impact on health in Poland persists, with practically no empirical base at this moment. Napiórkowska-Baryła and Witkowska-Dąbrowska (2018) initiated a public discussion with respect to Poland, arguing that energy poverty could affect the physical health of children suffering from reduced immunity or respiratory illnesses. There are also some first attempts by the Institute of Structural Research to assess the relationship between energy poverty and health in Polish coal regions; the results should be publicly available in 2020.

Notes

1 Based on the data provided by the Household Budget Survey (Statistics Poland, 2019).

2 Only 35 percent of all communities (around 800) took part in the survey, so the real level of energy poverty consciousness among local authorities could have been even lower.

3 The housing allowance is an instrument of Polish social policy, described as a monetary transfer to underprivileged owners or tenants in order to help them partly cover apartment maintenance expenditures. It does not help energy poor households much, since the average amount received in 2013 was about 200 PLN (less than 50 euro) per year per household and originally it was not supposed to cover energy costs and other expenditures, such as rent or water bills (Muzioł-Węcławowicz, 2015).

4 In Poland there are 16 regions called "voivodships."

References

Boardman, B. (1991) *Fuel Poverty: From Cold Homes To Affordable Warmth*. London: Belhaven Press.

Boguszewski, R. and Herudziński, T. (2018) "Ubóstwo energetyczne w Polsce." Pracownia Badań Społecznych SGGW, Warszawa.

Ciepiela, D. (2019) "Banki pomogą w walce o czyste powietrze," www.wnp.pl/energetyka.

Dworakowska, A. (2016) "Jakość powietrza w Polsce na tle Unii Europejskiej" in *Zrównoważone miasta. Życie w zdrowej atmosferze*. UN Global Compact Poland, Warszawa.

Eurostat (2018) "Energy," https://ec.europa.eu/eurostat/web/energy/data/database.

Figaszewska, I. (2009) "Ubóstwo energetyczne – co to jest?," *Biuletyn URE*, no. 5. Urząd Regulacji Energetyki, Warszawa, pp. 2–20.

Forum Energii (2019) "Czyste ciepło 2030. Strategia dla ciepłownictwa." Warszawa.

Herbst, I. (2013) "Potrzeby mieszkaniowe w Polsce. Dostęp do mieszkania w kontekście polityki prorodzinnej. Ekspertyza opracowana na potrzeby Zespołu ekspertów ds. wypracowania rekomendacji w zakresie polityki rodzinnej przy Kancelarii Prezydenta RP," Warszawa.

Hills, J. (2012) *Getting the Measure of Fuel Poverty. Final Report of the Fuel Poverty Review. CASE Report 72*. London: Department of Energy and Climate Change.

International Energy Agency (2017) "Energy policies of IEA countries: Poland 2016 review." Paris.

Jędrak, J. and Adamkiewicz, Ł. (2019) "Nierzetelne i wprowadzające w błąd informacje rozpowszechniane przez Izbę Gospodarczą Sprzedawców Węgla," https://smoglab.pl.

Krzykowski, M. (2014) "Ochrona odbiorców wrażliwych energii elektrycznej i paliw gazowych – uwarunkowania prawne," *Polityka energetyczna – Energy Policy Journal*, tome 17, vol. 3, pp. 257–268.

Kurowski, P. (2008) "Wydatki gospodarstw domowych na energię elektryczną. Próba ustalenia grup odbiorców wrażliwych na podwyżki cen energii (wyniki na podstawie badań GUS z 2006 r.)," *Biuletyn URE*, No. 5 (2008). Urząd Regulacji Energetyki, Warszawa, pp. 53–60.

Lewandowski, P., Kiełczewska, A. and Ziółkowska, K. (2018) "Zjawisko ubóstwa energetycznego w Polsce, w tym ze szczególnym uwzględnieniem zamieszkujących w domach jednorodzinnych," IBS Research Report 02/2018. Instytut Badań Strukturalnych, Warszawa.

Libor, G. (2019) "Ubóstwo energetyczne – śmiertelność i jej koszt. Szacunki na podstawie dostępnych danych." Izba Gospodarcza Sprzedawców Polskiego Węgla, Katowice.

Libor, G. and Bouzarovski, S. (2018) "The socio-demographic dimensions of energy poverty: a case study from Bytom, Poland," *Urban Development Issues*, vol. 57, no. 1, pp. 63–77.

Mędrzycki, R. and Szyrski, M. (2018) "Energy poverty as a European Union and Polish legal issue," *Białostockie Studia Prawnicze*, vol. 23, no. 2, pp. 125–138.

Ministry of Energy (2018) "Polityka energetyczna Polski do 2040 r. (EPP2040) – projekt," version 1.2, Warszawa.

Ministry of Finance and Ministry of Infrastructure and Development (2015) "Przegląd wydatków publicznych: Polityka wspierająca mieszkalnictwo," Warszawa.

Muzioł-Węcławowicz, A. (2015) "Problemy mieszkaniowe Polaków a polska polityka mieszkaniowa" in M. Salamon, A. Muzioł-Węcławowicz (eds) *Mieszkalnictwo w Polsce. Analiza wybranych obszarów polityki mieszkaniowej*. Habitat for Humanity Poland, Warszawa, pp. 13–92.

Municipal Centre for Social Aid in Cracow (2019) "Dopłaty do zwiększonych kosztów grzewczych," https://mops.krakow.pl.

Napiórkowska-Baryła, A. and Witkowska-Dąbrowska, M. (2018) "Fuel poverty – a new challenge for social policy?" *Olsztyn Economic Journal*, vol. 13, no. 1, pp. 45–56.

Official website of the City of Wrocław (2018) "Lokalny Program Osłonowy," www.wroclaw.pl/srodowisko/lokalny-program-oslonowy.

Ostańska, A. (2010) "Wpływ dotychczasowych termomodernizacji budynków mieszkalnych na oszczędność energii i planowanie programów rewitalizacji na przykładzie jednego z lubelskich osiedli," *Budownictwo i Architektura*, no. 7, pp. 89–103.

Polish Electricity Association (2018) "The contribution of the Polish energy sector to the implementation of global climate policy," Warszawa.

Polski Alarm Smogowy (2016) "Smog w Polsce, smog w Europie," www.polskialarmsmogowy.pl.

Polski Alarm Smogowy (2019a) "Fikcyjny Program Czyste Powietrze – PAS zapytał o realizację rządowego planu," www.polskialarmsmogowy.pl.

Polski Alarm Smogowy (2019b) " 'Kopciuchy' do likwidacji – kujawsko-pomorskie przyjmuje uchwałę antysmogową," www.polskialarmsmogowy.pl.

Pyka, M., Liszka, S., Czajkowski, J. and Kukla, M. (2014) "Ubóstwo energetyczne. Wyniki badania ankietowego oraz propozycje pomocy osobom ubogim energetycznie." Instytut na rzecz Ekorozwoju/Fundacja na rzecz Efektywnego Wykorzystania Energii, Warszawa.

Rolecki M (2019) "Europa marznie. O naszym ubóstwie energetycznym," Polityka, www.polityka.pl/tygodnikpolityka/rynek/1783416,1,europa-marznie-o-naszym-ubostwie-energetycznym.read

Rutkowski, J., Sałach, K., Szpor, A. and Ziółkowska (Święcicka), K. (2018) "How to reduce energy poverty in Poland?" IBS Policy Paper 01/2018. Instytut Badań Strukturalnych, Warszawa.

Sokołowski, J., Kiełczewska, A. and Lewandowski, P. (2019) "Defining and measuring energy poverty in Poland," IBS Research Report 01/2019. Instytut Badań Strukturalnych, Warszawa.

Sokołowski, J., Lewandowski P., Kiełczewska, A. and Bouzarovski S. (2019) "Measuring energy poverty in Poland with the Multidimensional Energy Poverty Index," IBS Working Paper 07/2019. Instytut Badań Strukturalnych, Warszawa.

Statistics Poland (2019) Household Budget Survey 2018.

Stępniak, A. and Tomaszewska-Kula A. (2013) "Ubóstwo energetyczne a efektywność energetyczna. Analiza problemu i rekomendacje." Instytut na Rzecz Ekorozwoju, Warszawa.

Supreme Audit Office (2019) "Efekty termomodernizacji wielorodzinnych budynków mieszkalnych będących w zasobach spółdzielni mieszkaniowych, realizowanej z udziałem środków publicznych," Warszawa.

Thomson, H., Snell, C. and Bouzarovski, S. (2017) "Health, well-being and energy poverty in Europe: a comparative study of 32 European countries," *International Journal of Environmental Research and Public Health*, vol. 14, no. 6, p. 584.

Węglarz, A., Kubalski, G. and Owczarek, D. (2014) "Propozycje mechanizmów wsparcia procesu przeciwdziałania zjawisku ubóstwa energetycznego w Polsce." Instytut na rzecz Ekorozwoju, Warszawa.

Wielgo, M. (2019) "Ulga termomodernizacyjna od 2019 r. Nawet 53 tys. zł odliczenia zakłada nowa ulga podatkowa," www.muratorplus.pl.

8 On how to fix a sturdy energy poverty system in Romania

Anca Sinea

A landscape of energy poverty

It was an unusually rainy and windy day at the end of January 2019, when I was driving through the streets of Bucharest with a large group of energy poverty experts from all over Europe to understand the energy poverty dilemma in this part of the continent. Observers could not help gazing at the mosaicked view of the patchwise-refurbished post-socialist tower apartment buildings that were flanking the streets on both sides. Some had learned about this sight from literature; for others the view was so familiar, as it is spread, with different intensities of course, all over the Eastern European landscape. Nonetheless, they were all bewildered at this striking depiction of a lack of a systematic approach and the failings of community enterprise to rehabilitate what were typically low-efficiency, fast-to-assemble panel buildings erected during communist Romania, which had found itself on a hype of artificially planned industrialization and urbanization against the clock.

Bucharest has exceeded EU average income per capita, being one of the fastest growing local economies in the bloc, and many other cities around the country have also become hubs of development and prosperity (World Bank Group, 2018). That said, this view repeats itself all too often across the country, pointing at fundamental causes and possible effects that remain untapped, untalked-about, silent. The few refurbishment programs implemented over the years that have managed to correct the city's aesthetics to a certain extent lack strategy, control, monitoring, and impact assessment, leaving their effectiveness uncertain. Concern about energy poverty has only met these projects incidentally, if ever. The programs currently in place are either void of funds or are administrative burdens for local authorities, which need to put additional effort into making tenants' associations coalesce, including making important financial contributions. Refurbishment projects in their current form will likely never target energy poor people, as needy households generally opt out due to their incapacity to invest (Primăria Sectorului Cinci al Municipiului Bucureşti, 2017). In the absence of a thorough evaluation of the housing sector to help focus these policies somewhat better, the only existing pictures of the state of the buildings are the incomplete and perhaps outdated 2011 census, the structurally alarming situations flagged by the World Bank, or maybe and sadly just the colorful view of a city stroll on an unusually rainy and windy day at the end of February. According to a 2018 World Bank diagnosis, over 3,100 residential blocks across the country would be in need of advanced and serious structural intervention in the event of an earthquake. These buildings house over 10,500 families. More than 60 percent of them are located in Bucharest (World Bank Group, 2018).

Striking to the eye, this colorful view was by far the most provoking circumstance of the moment. That particular day, the supply of heat and warm water was cut for over 250 apartment buildings in the Romanian capital city, leaving thousands of households utterly in the cold for several days (Ziare.com, 2019). This was not a mishap. Sources report over 3,500 such events every cold season with hundreds of thousands of families being forced into the cold repeatedly and even over several days and during Christmas, as was the case in 2018 (Kolbay, 2019).

The root cause is complex. What was hailed to be, and still is at times, one of the most efficient urban heating systems, built during the socialist era, with little investment afterwards, has grown outdated. Breakdowns in the Bucharest thermal system are ordinary. Over a million people learned to cope. This is the size of the population supplied by RADET, the largest thermal system in the country and second in the world only to that of Moscow (Nicut, 2010). RADET has over 40 percent of the local energy market (RADET, n.d.), for which 50 percent of energy costs are still subsidized (Ivanov, 2018). Households who are better off have installed electric boilers and heaters, the latter to be used during outages, because heating with electricity is much more expensive. They also turn the electric boilers on to compensate for the lag in hot water, because, they say, it takes a long time for warm water to start running through the pipes before they can really use the showers. If they do not use electric heaters, all the wasted water adds up in their bills, and this is what they aim to prevent. But if they do so frequently, the costs of household electricity can soar.

RADET, the local provider, lists all service outages on their website on a daily basis. But as it can barely cope with the system overhauls or with keeping track of failures, one organization came up with an online tool on a Google map interface where people can report system breakdowns in real time (CD, 2019). Whether this tool helps fix the system even in the short term is a matter of debate. What is sure is that Nicusor Dan, the person behind the initiative, a mathematician with important political involvement – having been the leader of one of the most dynamic parties of recent years – has put the topic of urban transformation in Bucharest on the political agenda. The two most visible topics are the state of the district heating and the need for structural intervention on severely affected buildings. Union Save Romania, initially Union Save Bucharest, has an important voice in the local General Council and a growing political position in the country, as it ranked third in the 2019 European Parliament elections. For decades, the district heating issue has been a constant point of tension in the local political debate along with other abounding structural issues in Bucharest. But whereas the state of buildings has been solely viewed as a matter of preventing yet another humanitarian disaster as in the case of the 1977 earthquake, the state of the district heating is presently a pressing one for individual and city budgets.

After years of oscillating between a political and civil society status, questionably mastering the skill to turn public concerns into consistent political agendas, Dan captured the political support of the entire opposition spectrum with a high chance of putting forward an agenda loaded with the heated topic of urban well-being for the upcoming autumn 2020 general elections. It is sure that these topics will monopolize a very complicated political debate. The Bucharest local elections have earned their reputation for being especially high stakes with difficult battles for the political parties, while politicians aim to bring forward topics of general interest. Given the context, how political forces will play into this issue a half year from now (with a coronavirus crisis in between) is quite difficult to anticipate, and the risk is high that the concern of energy poverty might sink under more general approaches, or succumb to more urgent matters of the day (i.e., COVID-19). The opportunity for more focused approaches coming from civil society, such as an initiative on increasing energy efficiency in the housing sector (EPG, Petrom, 2020) or one very specifically targeting energy poverty (UBB, 2020), to become visible in the debates will also be quite dependent on the detailed context.

However, irrespective of this uncertainty, the issue is administratively in the hands of the General Council (Parlamentul României, 2006), as a very strained issue in local politics, which has for decades failed to produce convenient solutions (Nicut, 2019a). Due to its many inefficiencies, RADET and its thermal agent producer, ELCEN, have accumulated debts over the years, rendering the entire system unable to invest in technologies. The state of insolvency means that both companies are unable to raise funds, including European funding, for their needs.

Many candidate solutions have already been tried out, from writing off at least part of the debt under various forms, to establishing new commercial entities under local administration. Currently the situation lies in legal uncertainty with no fix in view (Kolbay, 2019). It is uncertain for how many more years the system will be able to provide households with the necessary heat and warm water. At the moment, the generous subsidies (50 percent for all connected households irrespective of their situation) offered by the local administration not only shroud the enormous inefficiencies of the system, but they are also a part of the dispute. Not having been paid on time, subsidies accumulated into enormous arrears all along the supply chain. Subsequently, despite having committed to producing the necessary thermal agent within the required technical parameters (Nicut, 2019c), ELCEN threatened to cut the supply of warm water and heat for the population of Bucharest long before the former cold season had even set in (Nicut, 2019b). The population is caught between poor local administration and low supply capacity and no energy alternative but to use electricity at high costs, and only for those who are well enough off to have installed electric facilities for water and space heating. The other option is a very low standard of comfort.

Beyond just a mere matter of affordability, the district heating situation in Bucharest is an authentic problem of poor policy-making and public management with an impact on the population. Local systems across the country are affected by the same lack of investment. In Romania 1.3 million households are connected to the district heating in 60 cities around the country. But it would take up to 4 billion euros to invest in these capacities and a high degree of public management to optimize district heating (Ministerul Energiei, 2016). High losses and inefficiencies now result in massive consumer disconnections everywhere, leaving behind those who cannot afford to shift to individual boilers or retrofit their apartments. What is more, this unstoppable phenomenon causes an increasing burden particularly on these remaining households, who bear the shared costs of the system, along with its increasing failures, in a constantly decreasing crowd of district heating consumers.

But what is specific to Bucharest is the extent of the issue. Whereas elsewhere solutions are much easier to implement, and in some cities these have been found and have proved successful (Ministerul Energiei, 2016), in Bucharest the market size makes any quick fix utterly impossible, and political controversies with national implications are unavoidable. Some experts on the market say that the most intuitive alternative, connecting everyone to gas instead, is logistically a highly complex task that, at current market capacity (available construction services, costs and procedure involved, etc.), is only possible at a maximum rate of sixty thousand apartments every year. A complete transition to gas would consequently take up to ten years. Given the situation, such a schedule is unacceptable by any imaginable standard (Nicut, 2019a). Under the present circumstances, any solution for Bucharest will be of national importance, as it is bound to need vast resources, either through a national bailout or other financial solution. Certainly any way out will be beyond just the local budget and local competences (Nicut, 2019e). Irrespective of their individual material situation, and without minimizing the plights of those households that are the worst off, as the situation stands consumers of district heating are in a state of indefinite vulnerability.

The challenges of energy market liberalization

The thermal system in Bucharest and elsewhere in Romania and the state of the buildings are not the only two situations that are indirectly keeping the energy poverty issue high on the public agenda. Inefficiencies run deep in the system and the fact that a prospective negative national energy balance could bear down on private households is keeping the concern alive. Romania has for decades relied on diversified energy resources, most of them produced in national soviet-era plants that have been updated or developed only to a very small extent thereafter. Breakdowns in the national energy system are becoming a regular occurrence due in particular to the lack of capacity to balance out the system, with a direct impact on the market price and consumer bills. The summer of 2019 has been notable in that respect, as the media called out historical national peaks and even European records in energy prices and considerably more expensive imports due to the legal climate, the political context, and the state of the national energy system (Nicut, 2019f).

All things considered, in the larger context of market liberalization, to which the government has especially committed when adhering to the conditions of the European Union, the concern to keep energy prices affordable for households was constant over time in the country that has one of the poorest populations and one of the highest rates of inequality in the EU.[1] The deregulation of prices was negotiated with international financial institutions (Guvernul României, n.d.) and the European Union, in 2012 and 2015 respectively, and it resulted in a government commitment to a six-year-long steady calendar of energy market liberalization for private households, to be completed by 2021 for gas and 2022 for electricity. The process involved a constant increase of consumer prices at a rate of at least 10 percent per year (Capital, 2012). The nearness of the deadline fueled the political agenda. That agenda was already heated with mounting inefficiencies and the poor performance of politicians, with feelings of displeasure with energy and other multinational corporations regarding presumed under-taxation, resource exploitation, and hidden market intentions even while the population felt increasingly burdened and disenfranchised from its own resources. In 2018 this entire context led to the enactment of two pieces of legislation that were reportedly aimed to safeguard households as liberalization was completed. The topics outlined new forms of energy vulnerability present in society and resulted in highly controversial measures that were criticized for misinterpreting, mistargeting, and manipulating the concept of energy vulnerability and poverty.

Law no. 256/2018, or the so-called Offshore Law, was meant to establish a firm fiscal framework for deep-water energy exploitation and distribution in the context of newly identified gas resources in the Black Sea. Although heated, the public debate was brief and largely unilateral, excluding most of the stakeholders on the market. One of the provisions foresaw progressive taxation on supplementary revenues generated from gas price increases, whereas only a maximum of 30 percent could be reinvested and was, thus, exempted from taxation. The underlying argument was to use these additional revenues to develop the national gas distribution network in order to increase the energy security of the national population. Furthermore, holders of deep-water exploitation contracts were compelled to sell at least 50 percent of their production on the national market presumably to guarantee the necessary market liquidity at a "national" price for the planned infrastructure. The measures were met with great criticism from investors, who halted and postponed their projects to an uncertain future date, due to the conflict between these regulations and their market-based business model (Deacu, 2018). They were blamed for preventing the investment in the infrastructure needed to increase energy access on the national market (Nicut, 2019d). The measures taken triggered a set of dynamics that nullified the basic political assumptions. Some of them will be discussed below.

So far, in Romania household access to energy has generally been assessed inaccurately. With regards to gas, the National Energy Regulator[2] and the National Statistics Institute[3] give different accounts of the number of households that use gas. The most synthesized information on the topic is published in the draft national energy strategies (Ministerul Energiei, 2016), which points out that out of 7.5 million permanently inhabited households in Romania, only 2.5 million used gas for heating in 2015. That means that only approximately 33 percent of the households in Romania are able to use gas. But 44 percent have gas in their proximity. Gas mainly covers high-density urban or suburban areas, localities that are close to extraction sites and, therefore, are geographically more accessible. However, most unconnected administrative units are located at short distances from the network. For instance, 874 out of 2,304 unconnected administrative units in 2018 were located at less than 10 km away from the network (Sinea et al., 2018b). This evidence points to a number of limitations to access, which are not addressed by the Offshore Law.

First, there are high costs involved in connecting individual households to the gas network. Many households refrain from connecting to gas because they cannot afford the initial fees, which can rise up to 1,000 euros per household at an annual household income per capita of 4,500 euros in 2018 (CEIC, n.d.). In rural areas, which are the areas most affected, incomes are much lower than the statistical mean. To that, we might add the complicated application process that impedes households from pursuing connection (Sinea et al., 2018b). Several entities have raised these issues, recommending that the state subsidize connection fees as a protection measure for low-income consumers (EPG, 2018).

Second, the administrative process, as laid out in law (Parlamentul României, 2012) (ANRE, 2015), can be complicated not only for households but also for local administrations. Looking at the types of localities not connected to gas, it can be concluded that the smallest communities are affected in larger numbers. They also have the lowest capacity to go through the necessary procedures and elaborate projects and raise the necessary funds. The low associative capacity between smaller entities, which is a general trait of Romanian politics, may set further limits in facilitating energy access for households.

Third, in some situations it has been noted that communities are reluctant to connect to the grid and to commit to a long-term contract, opting for the more traditional resources, particularly fuelwood, which is usually self-procured, and might actually be cheaper than gas.

In rural areas, 82 percent of the population uses wood for heating. Having little alternative, these households are exposed to a number of vulnerabilities, such as substantial price variations. Apart from the households that use unconventional sources of wood and other combustion materials, it is important to note that in Romania the wood market is under-regulated and particularly volatile, with significant price differences between seasons and regions. The price of wood can soar especially during the cold season. Under normal circumstances and based on a standard room temperature, the price difference between wood and gas is slight, the first being comparatively cheaper per kWh. However, at peak moments the price of wood can easily surpass that of gas. Heating assistance benefits offered for wood are insufficient, while market intervention to protect consumers is impossible (Sinea et al., 2018a). The issue regarding the price of wood comes up every winter in the mass media. However, it is discussed primarily from a narrow industry perspective and less commonly with regards to the energy welfare of households. Wood consumption in areas with no access to gas and other heating resources brings up yet other dimensions of energy poverty, such as the quality of indoor and outdoor air associated with the low quality of stoves, inefficient consumption due to poorly insulated buildings, outdated appliances, and partial and temporary space-heating habits, which are solid traits of consumption behavior across the country (Sinea et al., 2018b).

Unlike gas, electricity is generally more widespread, with relatively few households remaining unconnected to the grid. However, uncertainty persists also on this topic as various accounts report anywhere between twenty thousand and one hundred thousand homes having no access to electricity, while the issue is typical of very-low-income households (Sinea et al., 2018a). Electricity is commonly used for lighting. Only 1 percent (Ministerul Energiei, 2016) of the population, mostly those who are very poor, uses electricity to produce heat.

The underlying rationale of the legal provision can be a matter of debate in itself: if companies do not follow through their investments under the law, there will be no outputs, no revenues, no taxes, no investments. This rationale notwithstanding, discussion regarding access to energy, and, more contextualized, the low territorial coverage of the gas distribution system, entails much more complex ramifications than any of the political arguments in favor of the so-called Offshore Law were ever able to display. Of course, the district heating crisis in Bucharest, with regard to the development of the gas network to connect households, can only display the possible large-scale deadlock in view, if an exclusively infrastructure-based approach were to be applied, based on the Offshore Law, assuming that the model would also work in practice. Many underlying aspects of energy poverty would be disregarded. In the end, in the context of a negative national commercial balance, questionable economic growth, and a diminishing curve of popular support, the government showed willingness to abandon the legal provision and restart discussions on the fiscal system without displaying any intention to approach the energy vulnerability topic any more closely.

A second and more heated debate was related to Government Ordinance (OUG) 114/2018 (Guvernul României, 2018) and the ensuing OUG 19/2019 (Guvernul României, 2019). In a nutshell, the ordinance was a collection of fiscal measures across topics aimed at increasing government revenues in a time of uncertainty related to public liquidity. The ordinance, which was also a matter of minimal public debate, was intentionally issued during the period of complete public shutdown between the winter holidays. Despite that, it quickly spawned heated debates rounding out all possible voices on the market. At the core of this legislation were two main systems: banking and energy. In anticipation of the bill, these economic sectors, which generated important foreign investment, became targets of political narratives and parliamentary investigations, thus provoking many conflicting emotions, much vehemence, and great uneasiness. In the public opinion, which is barely initiated on such technical issues, the puzzles were based on the easily grasped narrative of the foreign assailant that takes advantage of national resources giving little in return and is, hence, to blame for the lack of development and general social welfare. Given this rationale, the bill was aimed at settling the matter by additionally taxing the incomes made by these companies to eventually use these extra funds to fill in budgetary holes. But beyond this general approach, a number of provisions were flagged as groundbreaking in causing improvements in the field of energy welfare for the population. This seemed to be a two-track package aiming to tackle energy poverty through market forces and social benefits.

On one hand, the government hoped to freeze the remaining efforts needed to reach complete market liberalization, and which involved mainly price increases for household consumers. The measures were designed to target the entire population and not just consumers who were generally classified as energy vulnerable. OUG 114 introduced a cap on the wholesale price for gas destined for residential use and district heating systems, and the regulation of electricity prices for residential consumers, until February 2022. On the other hand, the ordinance also established changes to the existing benefits system to increase the number of beneficiaries by updating the income margins for qualification. While minimum salaries had been increased several times, the income requirements in this piece of legislation have not been updated accordingly. The social-democratic government of the time was highly criticized internally and by the European Commission for excessively intervening in the market to set off distorting effects that would eventually turn on the consumers, and for perpetuating economic inefficiencies that could otherwise be prevented if the energy market would be completely liberalized. Various voices pointed out that targeting these measures specifically to energy poor people would be add value to the performance of the energy market, while the current measures resembled a massive subvention. The entire situation eventually pushed the topic up the electoral agenda of all major political parties for the presidential elections of winter 2019.

In practice, the two-track intervention package prescribed was a weak policy that could only set off marginal and very-short-term changes of some kind to the state of affairs. It was essentially disconnected from the real challenges of energy poverty in Romania, as partly described above, while only managing to settle some deficits of energy welfare in a few households through cash distribution, something that could under no circumstance cause real improvement. The much-needed interventions at the grass roots level were for once postponed in favor of the well-established state of affairs.

The downsides of blanket measures and price-related energy poverty mechanisms, or the negative impact of regulated markets on the energy welfare of consumers, have been extensively discussed in the literature and are covered by a large collection of precedents. Therefore, further enhancement is unnecessary here. With regard to the existing relief system, some aspects need to be considered.

As one of the oldest and most institutionally encompassing systems, the Romanian energy poverty intervention scheme, or more correctly the "energy vulnerability scheme," is particularly extensive and established, wielding a great deal of decision-making power at the local level. Yet, its design hides a number of institutional and policy-related weaknesses that need to be explained, as they perpetuate many errors and conserve a cumbersome system in the face of a rapidly growing and more sophisticated phenomenon. The energy poverty intervention system is built on one of the oldest national policies – the social system – as one of the fundamental pillars of the Romanian welfare state. It is mainly a heating benefits program allocated distinctly for wood and other solid fuels, gas, and district heating, but only marginally for electricity with some additional elements of electricity subsidy for households that have a low consumption rate and very low income.

The institutional span of this policy is comprehensive, which not only grants the energy poverty intervention system a great deal of legitimacy but also makes it very complex. From Parliament to the government, who issue primary legislation, to the National Energy Regulator, which determines the technicalities of these rulings through secondary law, and the Labor and Energy Ministries, which set out the criteria; to the local authorities, who disseminate and process applications, and local agencies and distribution system operators (DSOs), who settle the payments: the energy vulnerability intervention tool is maneuvered by a number of entities (Parlamentul României, 2012) (Guvernul României, 2011). Any attempt to intervene in the system has to consider aligning a large number of entities. Certainly the hardliner mentality and resistance involved are typical of any public administration. This is aided by the prevailing lack of capacity and understanding of the matter. Ordinance 114 succeeded in bypassing the system while attempting to settle the energy poverty issues on the market (through one of its tracks), while through a second track – the welfare system – it only turned up one minor technical flaw: the provision did not touch upon the deep structure of the system.

The local branch of policy-making is particularly articulated. The allocation of heating benefits takes place by means of an institutional triangle that involves the County Offices of the Labour Ministry for social payments and affairs, the DSOs, and the town halls. The model is replicated at the level of every administrative unit (town or commune), which means that over three thousand local administrations are involved. The institutional design corresponds to the national administrative structure of the country, which makes it particularly stable. This leads to a situation where funds can reach all possible constituencies. But it is also a means of political validation, of which local politicians are well aware (Sinea et al., 2018a).

However, in this formula the political stakes of benefits or any type of aid can become quite high, as displayed by the situation of the Bucharest district heating subventions illustrated above. The situation has been confirmed in local surveys with regards to heating benefits, as respondents oftentimes relate benefits to mayor's efforts (Sinea et al., 2018a). Local authorities enjoy a lot of freedom in terms of implementation and they are well aware of the political advantage this financial tool may give them. This particular feature of the benefits system can produce two contrasting effects. On the one hand, the system is allowed some suppleness given the freedom of decision-makers to adapt to local needs and allocate resources as they find suitable. This may also result in incongruences in implementations, faults, and biases. On the other hand, the system is well embedded locally, and stable, so any attempt to improve it may result in resistance to change, and all prerequisites for reinforcement are there. Besides guaranteeing a large outreach, this may also conserve many of the faults and biases of the system indefinitely.

Practically speaking, local authorities decide on the strictness with which they apply the eligibility criteria, especially as in some respects they tend to be fairly vague. For instance, when the household budget is being analyzed to establish if the applicant's income is within the normative threshold, the admissions officers can evaluate and add up the applicant's possessions according to a model recommended by the Labour Ministry. However, as this is not a strict provision, there is great variation in implementation. Some decide to count goods strictly, some do so partly, and some not at all. The rest of the process happens through the effort of social counselors, who are aware of the situation in their respective constituency. But there are too few counselors to investigate each case objectively, and more often than not they act under the political guidance of the mayor. They are aware that the social grants are a powerful political validation instrument and, therefore, seek to operate them in ways that maximize gains (Sinea et al., 2018a).

Some local administrations use this gateway to implement forward-looking and innovative measures. But practice shows that anything that goes beyond the existing social system and its institutional inertia may be difficult to operate and may require additional policy-making and operation capacities that are not easy to find in local administrations. Based on national legislation, town halls can supplement national funds from local budgets and allocate them indiscriminately to other categories of energy vulnerabilities, as local needs may require (Guvernul României, 2011). Many times, creative solutions are very likely to conflict with other legal requirements and financial or capacity limitations that place them in complete deadlock. The city of Alexandria, in the southern part of Romania, is a region that struggles with complex situations of economic and energy poverty, but also with a high rate of absorption of heating benefits, showing either that the needs there are high, or that the system is particularly functional. There, the mayor intended to break with the traditional approach of giving out extra subsidies, and created an energy efficiency package. The local administration budgeted additional funds for the retrofitting of the houses of social aid recipients. The town hall purchased the necessary materials, whereas beneficiaries had to assume the costs for construction services. These could not be tendered by the local administration due to legal limitations. Capacity to identify alternative programs, such as European funds, in order to overcome these constraints was limited as is often the case with administrations of various sizes. Alexandria is a county capital city and medium-sized municipality. The program ultimately resulted in a large quantity of materials being purchased by the town hall and distributed to beneficiaries. But a large part of these materials remained unused for the purposes for which they were intended, because most of the beneficiaries could not afford to pay for construction work themselves. As a result of this plight, many chose to sell these materials for extra cash (Sinea et al., 2018a).

Similar examples with some variations are spread across the country, confirming that local decision-making mainly happens successfully within the constraints of the established welfare system. Despite creative ideas, conservative decision-making routines prevail as a manifestation of a former overly centralized state, where local management rarely has the resources to go beyond system confines and inertia. Moreover, even if the design of the system allows for important action at the grassroots level, this conflicts with the lack of know-how and lack of capacity to find alternative financing for local projects in a country that has scored well below its entitlements to European grants. These two traits bear important consequences in terms of energy poverty, to be specific, and overall welfare and development more generally. In the end, the legitimacy of local administration taking action in terms of energy poverty is not congruent with their capacity to do so.

Second, what was intended through OUG 114 was to increase the number of beneficiaries by correcting one of the failings of the existing system: that is, by indexing income thresholds to the current minimum income levels. This is what was identified before as being the track-two measure on energy poverty adopted by the respective emergency measure. Over time the legal minimum income limits have increased, whereas the income criteria have not been altered accordingly. As a consequence, many were excluded from these benefits on a technicality even if their situation had remained unchanged. The flaw was perpetuated for several years, without intervention. Welfare expenses diminished, with the cost of many losing their entitlements. Based on the new legal provision around 130,000 people would have been readmitted into the program. This was an important technical correction; however, the impact was quite minor compared to the list of inadvertences that allowed for systematic technical exclusions over the years. Notwithstanding their own limitations, conventional meters display the fact that the gap between beneficiaries of some kind of energy-poverty-related protection and the number of energy poor people at the national level can be quite large. Based on one of the indicators, in 2015 for instance, only 17 percent of the households in need received heating compensations (Table 8.1).

There are various factors that contribute to the existence of these gaps. The vague and inconsistently applied selection criteria can be an important factor. Examples prove that some of the rules are applied with excessive bureaucratic rigor, discouraging many from applying. In the fifth sector of Bucharest non-resident homeowners are required to provide documents from the authorities where they are permanent residents, which may turn the application process into a weeks- or maybe months-long process. Bureaucracy also tends to be generally recognized as complicated due to the lack of cooperation between institutions. Applicants must provide over and over again the same documents to every authority involved and at the beginning of every cold season, and this is particularly discouraging for many vulnerable people, who often lack the necessary skills to surf through institutions and rules. There is also much territoriality at the level of every institution, manifested in the way the criteria are implemented, as well as at the level of data circulation and coordination up and down the institutional circuit, which if less constricted could help close the gap between problem and solution even more effectively. This may give birth to yet another problematic topic, data management, which is not a priority in this chapter but can be a large part of the explanation for defective policy-making. Factors that can explain the gaps between energy poor households and those who are beneficiaries of interventions for energy poor people abound and many have been discussed before. But the truth is that the problem of energy poverty goes far beyond some selective conventional income or cost indicators, and may be as diverse as the access to alternative options of consumption, access to information and fair criteria, consumption behavior, etc. Some of these cannot even be quantified but they come up in media accounts and public debates again and again, without much of a reasonable resolution.

Table 8.1 Allocated heating benefits vs conventional indicators of energy poverty

Indicator	*2013*		*2014*		*2015*	
	Households in energy poverty according to indicator (% of the total households)	*Overlap between actual beneficiaries and those identified by indicator (%)*	*Households in energy poverty according to indicator (% of the total households)*	*Overlap between actual beneficiaries and those identified by indicator (%)*	*Households in energy poverty according to indicator (% of the total households)*	*Overlap between actual beneficiaries and those identified by indicator (%)*
Heating benefits	7.4	100	6	100	4.6	100
2M	11.9	14.86	19	33.33	12.10	17.39
LIHC	12.3	27.02	16.9	41.66	9.90	30.43
M/2	12.2	24.32	18.7	16.66	13.5	32.6

Source: Sinea et al., 2018a.

Conclusions

Energy poverty has increasingly become a front-page topic in Romania over the past three years, albeit always attached to other, much higher political stakes. The present chapter has recounted three such instances, the most important of all: the Offshore Law, Emergency Ordinance 114, and the Bucharest district heating scandal. At the time they were topics of heated debate that caused great political division while marking public debate long after that. The latter of the three is still ongoing. For the time being the district heating predicament remains clouded, revolving around the same issues of company debts, inefficiencies, and ongoing hazards. Political debates are rather exploiting the situation without putting forth any viable solutions. But fortunately the cold season is over soon and autumn is far away. As soon as it has set in, local elections will be just around the corner. However, with the coronavirus crisis perhaps taking over the debate, chances are low for a constructive public debate on the topic, let alone the plight of the most vulnerable households, which are not few in the capital of Romania.

The Offshore Law was eventually revised in September 2019, diluting the financial provisions so much that the argument on facilitating the access to affordable energy that lay at its core became utterly trivial (Pirvoiu, 2019). OUG 114 was the object of an ensuing accountability vote for the Liberal government in late December 2019 to discard the price caps and other energy market interventions (Barbulescu, 2019). The law was filed for contestation, which was subsequently approved, at the Romanian Constitutional Court due to procedural matters (DIGI, 2020). In the meantime, the law is on the Parliamentary agenda with no precise deadline in view, and the number of recipients for heating benefits has fallen so dramatically that at present there are close to no financial safeguards available for vulnerable households.

In all three cases it was obvious that the mismatches between the solutions identified and the existing energy poverty challenges were important, that the realities in the field are not the ones fueling the political debates. Moreover, because the topic is commonly associated with rather high political stakes, there is always the prospect that it will become a facilitator of various interests rather than an objective per se. Hence, there is a high risk of populism, especially as it touches upon long-established issues of concern in Romanian society. Even though these issues have cropped up time and again in the media, energy poverty is not an acknowledged concept. There may be some occasional references to it, but the only legally accepted term is "energy vulnerability," which is itself vague and poorly represented in implementation procedures. This is the reason why interventions are sporadic and rather improvised. Nevertheless, the underlying institutional infrastructure to deliver aid is well established. It offers the opportunity for stability, flexibility, a flowing policy-making process (based on impact analysis and policy feedback) and, not least, outreach to households.

With all its biases and political flaws, the importance of the welfare system in delivering support to energy poor households has not been sufficiently observed. However, three main aspects need to be fixed first, in order to allow the system to start delivering. First, there is need for a national approach. The 2012 legislated National Action Plan has yet to be implemented. When it comes into existence, it will have to reflect realities on the ground properly and engage a variety of stakeholders to be able to bring together as many informed (or data-based) arguments as possible. Second, the local level decision-makers need assistance to increase capacity to absorb the needs on the ground and to deliver effective programs including through cooperation. To this end, the national level should be able to remove administrative conflicts and facilitate creative initiatives at the local level. At this point, the dissemination of good practices would be a natural result. Last, effective policy-making will only be possible if the flow of communication up and down the institutional structure and horizontally between institutions is more functional. This includes the flow of data, procedures, and criteria. In the meantime, at the point where there is no clear direction in the area of energy poverty, and rather a policy void, one question remains between experts: should fuel cost benefits prevail or should the government prioritize efficiency interventions to contain energy poverty? Without any doubt, across international spectra there is a preference for the latter. Choosing one of the two, or making the appropriate transition from one to the other, cannot be decided without having delivered on these three important topics first.

Notes

1 Recent Eurostat statistics show that in Romania the purchasing power of the population was 57 percent of the EU average in 2016, whereas the distribution of wealth is the most unequal in the EU with a Gini coefficient of 37.4. Up until 2016 final energy prices in Romania were much lower that the EU average, for both gas and electricity, with the lowest price on gas (34 EUR/MWh) and the sixth lowest price on electricity (132 EUR/MWh) for household consumers.
2 The NER measures access down to the village level based on the distribution licenses registered.
3 The statistics office publishes information on Romanian localities connected to natural gas distribution networks, broken down per administrative unit.

References

ANRE. (2015, 07 09). Ordinul nr. 104/2015 pentru aprobarea Procedurii privind elaborarea studiului tehnico-economic în vederea realizării obiectivelor din sectorul gazelor naturale. M.Of. al României.
Barbulescu, O. (2019, 12 30). Modificarile la OUG 114 vor fi intarziate cel putin o luna PSD a contestat la CCR legea pe care Guvernul Orban si-a angajat raspunderea.

Capital. (2012, 06 29). ANRE a publicat calendarul de liberalizare a prețurilor reglementate la gaze și energie electrică. www.capital.ro/anre-a-publicat-calendarul-de-liberalizare-a-preturilor-reglementate-la-gaze-si-energie-electrica.html.

CD. (2019, 01 17). Harta interactivă a frigului în Capitală. Nicușor Dan le cere bucureștenilor să semnaleze problemele cu RADET-ul. HotNews.ro.

CEIC. (n.d.). *Romania Household Income per Capita*. Retrieved 03 02, 2020, from www.ceicdata.com: www.ceicdata.com/en/indicator/romania/annual-household-income-per-capita.

Deacu, E. (2018, 11 12). Legea offshore a fost promulgată. Cele mai importante prevederi Citeste mai mult: /adevarul.ro/economie/investitii/legea-offshore-fost-promulgata-1_5be9980bdf52022f75665607/index.html. *Adevarul*.

DIGI. (2020, 02 12). CCR: Modificările la OUG 114 prin angajarea răspunderii Guvernului sunt neconstituționale.

EPG. (2018). *Report: The Outlook for Natural Gas in Romania and Proposals for Its Value-Added Capitalization*. Bucharest.

EPG, Petrom. (2020). *Romania Eficienta*. (R. Eficienta, Producer). Retrieved 02 03, 2020, from www.romania-eficienta.ro: www.romania-eficienta.ro.

Guvernul României. (n.d.). Noul calendar de liberalizare a prețului la geze naturale din producția internă pentru consumatorii casnici.

Guvernul României. (2011). HOTĂRÂRE Nr. 920/2011 din 21 septembrie 2011 pentru aprobarea Normelor metodologice de aplicare a prevederilor Ordonanței de urgență a Guvernului nr. 70/2011 privind măsurile de protecție socială în perioada sezonului rece. Bucharest.

Guvernul României. (2018, 12 28). ORDONANȚĂ DE URGENȚĂ Nr. 114/2018 din 28 decembrie 2018 privind instituirea unor măsuri i^n domeniul investițiilor publice și a unor măsuri fiscal-bugetare, modificarea și completarea unor acte normative și prorogarea unor termene. *Monitorul Oficial NR. 1116 din 29 decembrie 2018*.

Guvernul României. (2019, 03 29). ORDONANȚĂ DE URGENȚĂ Nr. 19/2019 din 29 martie 2019 pentru modificarea și completarea unor acte normative. *Monitorul Oficial NR. 245 din 29 martie 2019*.

Ivanov, C. (2018, 08 21). Radiografia dezastrului de la RADET: pierderi de 28,61% în 2017, circa 80 milioane lei, și peste 7.500 de incidente și avarii. HotNews.ro.

Kolbay, G. (2019, 2 21). Falimentul RADET este iminent. Ce solutii sunt pentru ca bucurestenii sa nu ramana fara apa calda si caldura – UPDATE. Ziare.com.

Ministerul Energiei. (2016, 11 15). Strategia Energetică a Roma^niei 2016–2030, cu perspectiva anului 2050 Versiune preliminară supusă consultării publice.

Nicut, M. (2010, 10 9). Sistemul de termoficare al Bucureștiului a ajuns să se autodistrugă, în fiecare oră în care livrează căldură. Explicații.

Nicut, M. (2019a, 08 20). Stab, Engie: Dacă toți bucureștenii ar renunța la RADET, am putea racorda toate centralele de apartament la rețea în zece ani. Nu e realist. E-nergia.ro.

Nicut, M. (2019b, 09 4). Elcen intenționează să oprească furnizarea apei calde în București, din cauza datoriilor Primăriei. *E-nergia.ro*.

Nicut, M. (2019c, 09 12). Elcen: Vom livra agent termic bucureștenilor fără sincope la iarnă, deși rețeaua RADET ne obligă să funcționăm sub minimul tehnologic.

Nicut, M. (2019d, 09 23). FPPG salută revizuirile propuse de Ministerul Energiei pentru legea Offshore și legea Petrolului. E-nergia.ro.

Nicut, M. (2019e, 09 25). Firea vrea OUG de la Guvern să poată plăti facturile către Elcen, sindicaliştii din centrale o acuză că toacă banii termoficării pe vouchere şi apartamente la suprapreţ. E-nergia.ro.

Nicut, M. (2019f, 09 27). România a fost foarte aproape de blackout, de două ori în acest an. E-nergia.ro.

Parlamentul României. (2006, 07 14). Legea serviciului public de alimentare cu energie termica. *Monitorul Oficial al României, 174* (XVIII) (Nr.651), pp. 1–13.

Parlamentul României. (2012). Legea nr. 123 din 10 iulie 2012 a energiei electrice şi a gazelor naturale publicată în Monitorul Oficial al României nr. 485 din 16 iulie 2012.

Pirvoiu, C. (2019, 09 20). Guvernul schimbă Legea Offshore în favoarea investitorilor din Marea Neagră- proiect de lege. Hotnews.

Primăria Sectorului Cinci al Municipiului Bucureşti. (2017). *Regenerare Urbana Ferntari*. Bucharest.

Radet. (n.d.). About. (Serviciul Sisteme Informatice – RADET Bucureşti) Retrieved 02 01, 2020, from www.radet.ro: www.radet.ro/despre-radet-bucuresti.php.

Sinea, A., Murafa, C., and Jiglau, G. (2018a). *Energy Poverty and the Vulnerable Consumer in Romania and in Europe*. Cluj-Napoca: Cluj University Press.

Sinea, A., Murafa, C., and Jiglau, G. (2018b). *The Natural Gas Opportunity in Romania's Residential Sector*. Cluj-Napoca: Center for the Study of Democracy.

UBB, a. (2020). *EnPower*. Retrieved 03 12, 2020, from www.energy-poverty.eu.

World Bank Group. (2018). *From Uneven Growth to Inclusive Development. Romania's Path to Shared Prosperity*. World Bank Group. Washington, DC: World Bank Group.

Ziare.com. (2019, 02 20). Ziua si avaria la RADET: Peste 250 de blocuri din sectoarele 2 si 6 sunt afectate.

Part III

Regional variations of energy poverty

9 Hidden energy poverty

The case of the Czech Republic

Hedvika Koďousková and Lukáš Lehotský

Introduction

The Czech Republic seems to be little affected by energy poverty. When compared to most other EU member states, the overall prevalence of energy poverty in the country seems to be low as measured by both consensual and expenditure-based indicators. National means for the "inability to keep home adequately warm" indicator showed a decreasing tendency from 6.7 percent in 2012 to 3.8 percent in 2016, levels very close to Germany, for example. Three percent of Czech households reported having "arrears on utility bills" in 2016 (down from 4.1 percent in 2012), which is one of the lowest scores inside the EU and comparable to countries such as Denmark, Germany, and Sweden. With respect to expenditure-based indicators, the "proportion of population whose share of energy expenditure in income is more than twice the national median share or the 2M indicator" of about 10.7 percent for 2010 was again more favorable for the Czech Republic than for most other EU member states (Thomson and Bouzarovski, 2018, pp. 23–28). The Czech "Low Income High Costs or LIHC indicator" was around 9 percent in 2012, slightly below that of the United Kingdom (Bouzarovski and Tirado Herrero, 2017, pp. 34–36).

Although the comparison of the Czech situation with the rest of the EU is quite favorable, this should not lead us to conclude that energy poverty is a minor problem in the country. Our thesis is that, despite good general results, specific strata of the population face energy deprivation to a larger extent but risk being systematically overlooked by the general indicators. Our chapter is structured around this argument in the following way. In the first part, we present the main characteristics of energy poverty in the Czech Republic. Here, existing knowledge is extended by use of local data of the Czech Statistical Office (CSO). Sociological research on peripheral areas is further used to explain the spatial distribution of energy poverty in the country. In the second part, we present the three types of policies designed to address the vulnerabilities identified.

Country-specific characteristics of energy poverty

When describing energy poverty in the Czech Republic, it makes sense to go beyond the general statistics mentioned before. A strong income dimension and a specific socio-demographic character of energy poverty in the Czech Republic have been confirmed by energy poverty studies (Bouzarovski and Tirado Herrero, 2017; Karásek and Pojar, 2018), as well as by Czech statistical data (Červenka and Ďurďovič, 2018; Czech Statistical Office, 2019). Several patterns can be identified.

First and foremost, in the Czech Republic there are significant differences between the figures for the entire population and those for specific income and socio-demographic groups. Although the share of housing costs in households' net income has gradually decreased over the past five years, as household incomes grew faster on average than housing expenditures,[1] there are important differences in these patterns across different types of households (Czech Statistical Office, 2019a). For instance, housing costs make up 36.3 percent of the net income of unemployed people; 29.9 percent of one-person households older than 65 years; 27.9 percent of the lowest 10 percent income decile; 27.1 percent of one-person household with dependent children; and 21.8 percent of the net income of pensioners. All of these percentages are well above the national average of 15.9 percent for 2018 (Czech Statistical Office, 2019b, 2019c, 2019d).

With respect to subjective feelings, generally speaking in 2018 almost 19 percent of Czech households considered housing costs to be a heavy burden. This perception is much more prevalent for some vulnerable households: 51.5 percent for unemployed people; 39.8 percent for the lowest 10 percent income decile; and 37.5 percent for one-person households with dependent children (Czech Statistical Office, 2019e, 2019f, 2019g). The above-mentioned groups also more often declared having "arrears on paying heating, electricity, gas and water in last 12 months," at 9.1 percent, 6.1 percent, and 4.7 percent, respectively, compared to just 1.8 percent of the total households in 2018 (Czech Statistical Office, 2019c, 2019f, 2019g). With respect to energy services, the latest survey done by the Public Opinion Research Centre (under the Institute of Sociology of the Czech Academy of Sciences) indicated that around 25 percent of respondents pay for electricity, heating, and water with "lesser" difficulties, while 5 percent do so with "higher" difficulties. Again, the study has shown that levels of difficulties correlate with household income and with the number of household members. One-person households, especially those of widowed or divorced respondents with dependent children, show a higher than average level of difficulty (Červenka and Ďurďovič, 2018).

Czech pensioners are generally perceived as a distinctive category when speaking about energy. Among pensioners, the percentage of households that consider housing costs a heavy burden is higher than the national average of 19 percent noted above, according to 2018 data (26.5 percent in the category of one-person households over 65 and 21.7 percent of pensioners). However only 0.9 percent of one-person households over 65 and 1.3 percent of pensioners declared "having arrears on paying heating, electricity, gas and water" in 2018 (Czech Statistical Office, 2019g), which is less than the national average of 1.8 percent noted above. These findings are in accordance with previous studies which suggest that people in Central and Eastern Europe put great emphasis on paying their energy bills on time and that energy debts are associated with some kind of negative social stigma (Filčák and Živčič, 2017, pp. 45–46). This is most probably why on average only a small percentage of Czech households declare arrears not only on paying for heating and electricity, but also for rents, mortgages, and dwelling-related loans – although a higher percentage of applicable households declared having arrears on paying other loans (Czech Statistical Office, 2019e).[2] The latter is more consistent with rather unfavorable statistics on enforcement proceedings in the country (Ecumenical Academy Prague, 2019).[3] Households formed by pensioners most probably prioritize housing expenses over other important expenditures such as food, health care, or clothing, and/or try to solve the situation by making strict energy savings by holding back on heating or electricity consumption.

Finally, a gender component can be identified with respect to energy poverty in the Czech Republic, both in general (Bouzarovski and Tirado Herrero, 2017, p. 39) and in the case of pensioners. According to a press release of the CSO, almost a tenth of Czech households (approx. 440,000) pay more than 40 percent of their disposable income for housing and energy. Almost two-thirds of them (approx. 285,000) live in rented apartments and almost three-quarters of them (approx. 205,000) are female pensioners living alone (Czech Statistical Office, 2019a).

Besides the strong income and socio-demographic character mentioned above, energy poverty in the Czech Republic is also related to technical characteristics of the building stock. In their study, Karásek and Pojar (2018) created a scaled-down database of buildings and households that helps characterize the current situation of energy poverty in the Czech Republic. They identified single-income families living in single-family houses as the most vulnerable type of household. A single-family house, since it does not share its perimeter walls with another heated space, has high heat losses if not retrofitted. In the Czech Republic, 76 percent of the buildings were built before 1990. Family houses built during the communist regime often fall short of the energy efficiency standards typical for houses built after 1990, having higher energy losses and higher energy costs, but also a large potential for energy savings.[4] This is why low-income households have to spend large amounts of their income on heating (Karásek *et al.*, 2016, pp. 29–30; Karásek and Pojar 2018, p. 134).

The findings of special survey on energy consumption in Czech households, conducted by the CSO in 2015, are consistent with these findings. As far as the characteristics of the building stock are concerned, 72.2 percent of all apartments (either in family houses or in apartment buildings) which have not been insulated (neither the walls, nor the roofs or through the installation of thermally insulated windows) were built before 1970; 13.2 percent between 1970 and 1980; and 14.6 percent after 1980 (Czech Statistical Office 2017, pp. 32–33). Of these uninsulated apartments, 34.1 percent are occupied by families with the lowest incomes (no more than CZK 20,000); 22.7 percent by households with incomes up to CZK 26,000; and the rest by three groups of households with higher incomes, i.e. up to CZK 33,000, up to CZK 45,000, and more than CZK 45,000 (Czech Statistical Office, 2017, p. 94). This means that of all apartments in the Czech Republic that have not yet been insulated by any means, about one-third are inhabited by the lowest-income group and more than a half by the two lowest-income groups. A closer look at low-income groups reveals that those living in apartments in single-family homes are much more likely to have a home with no insulation than low-income families living in apartment buildings. In particular, whereas 37.5 percent of households with the lowest incomes living in single-family homes have no insulation,[5] just 8.6 percent of households with the lowest incomes inhabiting apartments in apartment buildings have dwellings that have not yet been insulated by any means (Czech Statistical Office, 2017, p. 94).[6] One possible explanation for this is that the potential of the lowest-income families to spare sufficient money for refurbishing a single-family home is rather limited. Ordinary loans from financial institutions may be unavailable as well due to low incomes. One possible solution to provide refurbishment funds is state subsidies. However, as explained in the second part of this chapter, those who are below a certain income threshold have almost no chance of accessing a grant for efficiency measures, due to the need to provide co-financing and, most importantly, due to the system of reverse financing, by which the beneficiary is reimbursed after the refurbishment has been finished. This can represent an insurmountable barrier for families with low incomes or for other disadvantaged groups such as individuals with low education or a serious health condition (see below). To sum up, taking into account the technical characteristics of the building stock, the priority should be the lowest-income households living in old uninsulated single-family houses, where the most unfavorable ratio of low income to high energy costs is expected.

Last, given the evidence presented, it is important to discuss the spatial distribution of energy poverty in the Czech Republic. In their article, Bouzarovski and Tirado Herrero (2017) identify areas of low and medium population density as those hosting above-average numbers of households affected by energy poverty, measured by "energy burden" and "LIHC" indicators. The former reveals a concentric pattern (focusing on Prague) of the share of households experiencing energy burden above 20 percent, with the thinly populated Zlín region ranking the highest. In LIHC terms, the Zlín region is also understood as the most vulnerable, albeit the two neighboring

Northern Moravia regions as well as the far northwest Karlovy Vary region were also identified as relatively susceptible to the condition. No data is presented for the Vysočina and the South Moravian regions in the Bouzarovski and Tirado Herrero analysis (Bouzarovski and Tirado Herrero, 2017, pp. 42–44). The authors explain the vulnerability of the Zlín region by the prevalence of small and medium-sized settlements, where the highest proportions of households with high energy burdens and costs can be found[7] and where people have been marginalized in housing refurbishment programs (Bouzarovski and Tirado Herrero, 2017, p. 46, see also below). Although these findings are largely consistent with the above-described patterns of energy poverty, we believe that the spatial distribution of the issue requires further understanding.

If energy poverty in the Czech Republic mostly affects households disadvantaged by their income and socio-demographic characteristics, as well as low-income households living in single-family houses with respect to housing characteristics, Czech sociological research provides better insights into the geographical distribution of energy poverty. In particular, a number of studies have been dedicated to the identification and spatial delimitation of peripheral areas with high concentrations of social disadvantages that have negatively influenced the situation in these regions, along with analyses of contributing factors (Musil and Müller, 2008; Novák and Netrdová, 2011; Bernard and Šimon, 2017).

Two basic processes have been identified as affecting the concentration of social disadvantages in problematic areas. One of them highlights the impact of long-term changes in the geographic organization of society, which has led to the creation of new centers and thus to the decline of the importance of regions and settlements that have lost the functions of a center. One of the consequences of this process is the long-term economic stagnation and depopulation of hard-to-reach rural areas. The second process is economic restructuring after the fall of communism related to post-industrialization and market liberalization. Economic restructuring mainly hit regions with disadvantageous economic specialization, including both traditionally heavily industrialized regions and regions with a strong agricultural basis. The variety of processes and the diversity of the territorial development potentials can cause the emergence of different types of peripheries (Hampl, 2007; Hruška, 2014; Hampl and Marada, 2015, as cited in Bernard and Šimon, 2017, p. 7).

In their study, Novák and Netrdová (2011) confirm the existence of two categories of problematic regions. While the North Bohemia industrial area can be classified as structurally affected, the Bohemian-Moravian Highlands (*Českomoravská vrchovina*) in Vysočina region and other "non-developing areas" largely present in the northeast and northwest of the country, the Zlín region, and the southern part of the South Moravian region, are typically rural (Novák and Netrdová, 2011, p. 739). Although some characteristics are rather common for these territories, such as a high unemployment rate and a high share of households receiving housing allowances, some distinctive aspects have been identified. For example, whereas the North Bohemia region has a younger population and relatively high entrepreneurial activity, the Bohemian-Moravian Highlands suffers from a very low share of entrepreneurs and an outflow especially of the younger more-educated generation. This trend is also related to low housing construction intensity and a lack of job opportunities (Novák and Netrdová, 2011, pp. 736–737). Bernard and Šimon (2017) go further in this sense, and develop an approach that allows the delimitation of multiple types of peripheral areas based on four separate prevailing dimensions of disadvantages.[8] In their study, the northeast and northwest regions (together with some other areas near the frontier with neighboring states in the south) are defined as peripheries typically with high unemployment and social exclusion, a high share of households receiving housing allowances, and low education. On the other hand, e.g. on the border of the Vysočina region with neighboring regions, there are peripheries facing demographic challenges based on a high concentration of inhabitants over 65, a low share of the population younger than 15, low population growth, and a high share of inhabitants receiving allowances. The overlap of these two types of peripheries is quite weak. Finally, there are more dispersed areas in the Czech Republic, where the specificities of the Czech rural areas are most evident, such as a high share of people with education beyond minimal national requirements, a high share of inhabitants working in the agriculture sector with lower salaries, etc. (Bernard and Šimon, 2017).

To summarize, we see very diverse social-economic processes which have led to the creation of peripheries with multiple types of social disadvantages. Most of the problematic areas are largely consistent with the spatial distribution of energy poverty identified in the study of Bouzarovski and Tirado Herrero (2017), at least as far as the LIHC indicator is concerned. The only exceptions are the Vysočina and South Moravian regions, where no data were available in Bouzarovski and Tirado Herrero's (2017) analysis, many parts of which were, however, identified as problematic in both Novák and Netrdová's (2011) and Bernard and Šimon's (2017) research.

With respect to the spatial distribution of energy poverty in the Czech Republic, the regional database of the CSO also provides some guidance. During the last five years, the Karlovarský, Ústecký, and Liberecký regions located in the northeast of the country, the Vysočina region, and the Moravskoslezský region located in the northwest of the country have seen the highest percentage of households who declared either having problems with their dwelling (due to dampness); arrears on paying heating, electricity, gas, and water in the last 12 months; or considered housing costs a heavy burden (authors' own calculation based on data from Czech Statistical Office; for latest see Czech Statistical Office, 2019h, 2019i). Whereas problems with dwellings were present mainly in the Vysočina region, which

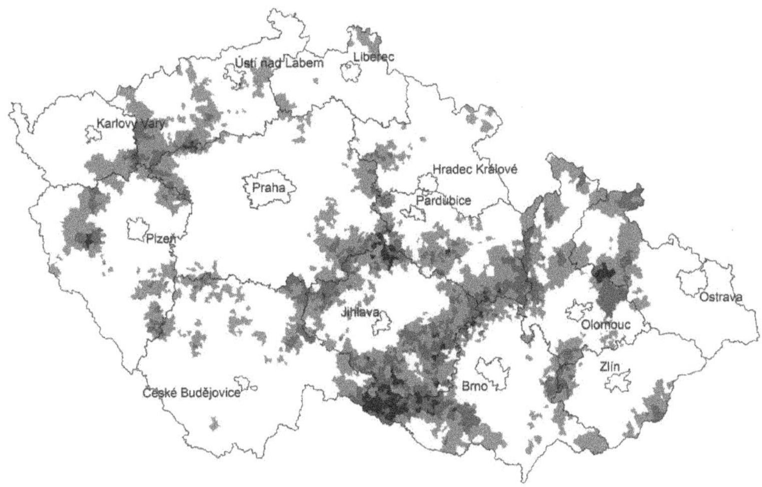

Figure 9.1 Regional decline in the Czech Republic.

Source: Novák and Netrdová (2011, p. 734).

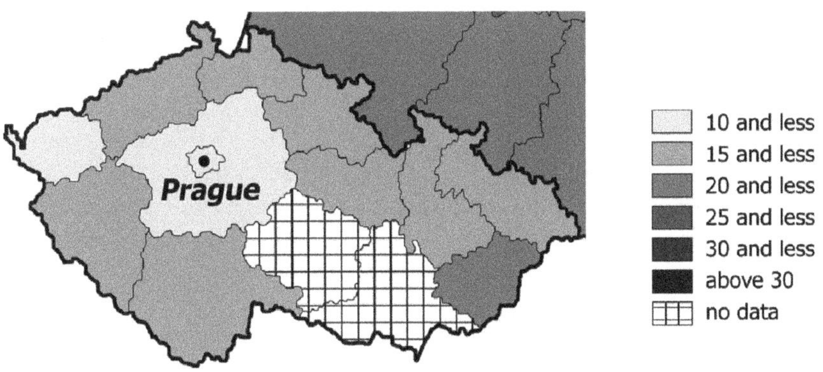

Figure 9.2 Regional variation in the shares of households who are energy poor according to the LIHC indicator.

Source: Bouzarovski and Tirado Herrero (2017, p. 44).

points to the poorer state of the housing sector in the region, households in the Ústecký and Moravskoslezský regions more often declared having arrears on paying for energy services and considered housing costs a heavy burden, which is consistent with the map of the highest concentration of energy poverty.[9]

Overall, the above-mentioned data suggests that energy poverty in the Czech Republic is somehow "hidden," as the most affected households are hard to find and survey. The presence of energy poverty in the country is not so obvious from the conventional indicators for the entire population, as it affects individual households which are scattered, but still forming a community based on their income and socio-demographic characteristics. Its geographical occurrence is uneven, as it is localized in certain areas or places, where various socio-economic processes led to higher concentrations of multiple forms of social disadvantages.

Energy poverty as a part of the wider question of housing

Given the situation described above, there are generally two types of policy areas that are fully capable of addressing the issue of energy poverty in the Czech Republic. However, our assessment is that both have somehow mistargeted the population most affected by energy poverty and from this perspective they do not live up to their full potential. The first type of policy is related to the housing needs of the population, and accessibility and affordability of accommodation for low-income households.

Legacies of the current situation can be traced back to the communist era of 1948–1989. Before 1989, the state owned the vast majority of the housing stock, mostly in the form of new condominium houses built between the 1950s and the end of the 1980s. The houses built by the state remained under state ownership, but the state granted individuals an indefinite right to use a unit in the house. Moreover, some flats were allocated to housing cooperatives at the municipal level. These organizations were also granted an indefinite right to use the property. The housing issue and construction of new facilities was a major issue during the entire communist period – the construction speed could not match the ever-increasing demand for housing. Nevertheless, large-scale housing projects were undertaken especially in the cities during the 1970s and 1980s (Valentová, 2005).

After 1989 a new legal system was developed whereby houses were transferred into the ownership of either housing cooperatives, municipalities, or individual holders. This massive privatization process, which was primarily managed by municipalities, led to a diversity of forms of ownership on the market (Valentová, 2005), which was also identified as a source of new social inequalities (Sunega and Lux, 2018). Inequalities grew quickly after the end of communism. More in-depth studies claim this is the primary reason why Czechs seem to strive to own their houses, as house ownership is a matter of prestige (Sunega and Lux, 2018). Housing statistics illustrate the high ownership rate, with only approximately 22 percent of the dwellings rented – well below the EU average (Ministry of Regional Development, 2018).

Looking into the situation of these renters is relevant to our discussion as they are mainly people who could not afford to own the houses they live in. Moreover, in their situation they are affected to a large extent by the fluctuations of prices on the housing market. During the 1990s rents were still regulated at a low rate. The regulation applied predominantly to dwellings that were still under municipal ownership. However, when deregulation took place in 2003, houses that had previously been three to four times cheaper than unregulated ones converged with the market (Mikeszová *et al.*, 2009). As a result, some of the occupants of rented flats were unable to pay the new prices and had to move to smaller flats, or, as happened in some cases, had to evacuate standard housing (Hruška *et al.*, 2013).

The Czech state approaches the housing affordability question primarily through national social welfare programs (Valentová, 2005; Karásek and Pojar, 2018). There are two types of housing benefits – "the supplement for housing" (in Czech, *Doplatek na bydlení*) and "the housing allowance" (in Czech *Příspěvek na bydlení*). The first one is reserved for people in a severe state of material deprivation, who are vulnerable due to housing costs (Ministry of Labour and Social Affairs, 2019a). It is generally provided to people who cannot afford to pay rent in a dwelling and, therefore, live in non-residential spaces or in shelters. The second type of support (housing allowance) is provided to a larger category of people, those who have a low net income but are not directly threatened by poverty. It is available to owners as well as tenants who have a permanent residence and have housing expenditures above 30 percent (or 35 percent in Prague) of the median net income of all households in the building (Ministry of Labour and Social Affairs, 2019b). The housing allowance may be used for covering expenses on rent, but also energy (electricity, gas, etc.) and heat, as well as water. Thus, it is expected to address poverty in its larger interpretation, including energy poverty (European Commission, 2019). Due to the relatively generous social support, energy poverty is not manifested at this level. The interviews conducted by the authors highlighted the mitigating effect of the housing allowance on energy poverty.

Yet there is a significant part of the population that lives in dire conditions in either the so-called socially excluded localities or in shelters. The reason for their marginalization is either the lack of social housing or their high debts, a situation which literally excludes them from the housing market. These individuals are subject to illicit housing practices.

Some type of social housing policy and state-driven apartment development has been in place since the mid-1990s. But due to regulated rents, it made economically no sense to build social housing. At the time of deregulation the state supported the development of affordable housing through financial grants and low-interest loans for housing cooperatives and municipalities. As a result, the concept of social housing was primarily understood as a measure to increase affordability and widen the housing ownership of the lower-middle class, not as a measure to address the availability of housing to the most vulnerable strata of the population (Valentová, 2005). Moreover, the social housing issue is solely confined to municipalities, nongovernmental organizations (NGOs), or other non-state actors

(Valentová, 2005), while as of 2019 no nationwide policy or legal framework focusing on social housing was in place. Due to that, there is no clear framework setting out rules for providers and users of social housing, nor responsibilities for any of the parties involved. The national regulation only addresses the general tenant–landlord relationships (Gabal and Víšek, 2010, p. 20; Prokop, 2019, p. 447).

It is fair to say there were attempts to adopt legislation regarding these issues. The most recent attempt was in 2015 with a draft legislation on social housing. The document, which was prepared by the Ministry of Labour and Social Affairs, provided the much-needed national legal framework. Unfortunately, it was delayed in the procedural cycle and the Czech government approved it only in 2017 (Government of the Czech Republic, 2017). That being the last year of political mandate for the government, the law did not have enough time to pass through all legislative procedures in the Czech Parliament and was not adopted. The issue was sidelined by the subsequent government.

Along with the lack of social housing, the issue of debts is yet another factor that forces people out of standard housing into non-residential spaces, shelters, or even into homelessness, subsequently preventing them from regaining their houses. Most often these are debts on rents or debts on common services associated with housing, such as wastewater treatment fees, communal waste collection fees, and energy. Energy debts were singled out as a significant driver of homelessness in one of the interviews conducted by the authors. Existing research on debts points in the same direction (Hruška *et al.*, 2013). People accumulate debts and once these go above a certain level it becomes nearly impossible for the affected people to leave the so-called "debt trap" (Szénássy *et al.*, 2017). Indebtedness prevents these people from finding regular housing. Landlords typically require a deposit of several rent installments in advance, which discriminates strongly against indebted people. Furthermore, energy companies refuse to sign a contract before all debts are repaid. At the same time, municipalities owning municipal housing require that a person proves there are no overdue debts or distrainment procedures taken against her (Foldynová *et al.*, 2015; Szénássy *et al.*, 2017). One of the respondents interviewed also highlighted debts on water and sewage rates as a limitation. For debtors with multiple overdue debts, these are insurmountable barriers to adequate living standards. With inadequate heating during the winter months (most often using energy-intensive and inefficient electric radiators), lodgers are exposed to the vicious circle of poverty, debt, insecurity, expulsions, and relocations.

The lack of social housing for the most vulnerable people as well as the absence of rehabilitation mechanisms for severely indebted people and the "supplement for housing" for people in material need created an opening for the emergence of so-called "poverty entrepreneurship." Housing for deprived people is provided in private lodging houses divided into several apartments.

These houses lack any investment and are often well below legal housing standards from multiple points of view (hygiene, safety, energy efficiency, etc.) (Szénássy *et al.*, 2017). "Poverty entrepreneurs" target vulnerable individuals who are in a very complicated life situation without access to the standard housing market. In many cases, these lodgings offer the only option for housing and the only alternative to homelessness. Thus, people in dire situations have little choice and accept any dwelling available (Hruška *et al.*, 2013). In 2015, 70,000 people lived in lodging houses and an additional 110,000 were at risk of losing their standard home (Prokop, 2019, p. 448).

Lodging houses are a consequence of the retreat of the state from the housing sector. Their conditions can be a source of vulnerability rather than of support. First, the price of the rent is calculated based on the assumption that the lodger receives the supplement for housing and it is often set well above the level of the supplement. Second, dwellings are often provided without a proper long-term tenancy contract. Thus, the legal protection for tenants is tenuous, despite the fact that Czech housing law is generally favorable to tenants. Based on the contracts, when renters experience difficulties with payments, most often they are forced to evacuate the apartments immediately without prior notice and with no time to look for other opportunities. Rent contracts often contain harsh clauses on evacuation. Most often, these stipulations are not in contradiction with national legislation, which is general and vague on these topics. However, poverty entrepreneurs assume that renters lack both the access to legal consultancy and understanding of legal nuances to fight back. They also take advantage of the fact that housing alternatives are limited.

The living conditions offered by lodging houses are below optimal. Investments are scarce and lodging houses are in poor condition, and the access to reliable energy services is limited. Given the legality of this housing model, there are very few options for the state to address its structure. There have been some attempts to limit the phenomenon, but the effects of these attempts have burdened the lodgers themselves, generating potential to worsen the situation. A vivid example of this is the 2016 amendment to the social welfare law, which allowed municipalities to identify areas with a higher concentration of "socially unacceptable phenomena." These areas are usually associated with higher crime rates, higher occurrence of drug addictions, and other social deviances. The areas thus risked being designated as "no-benefit zones" (Zákon č. 111/2006 Sb.: Zákon o pomoci v hmotné nouzi 2019), so that anyone living in such areas would be denied access to the supplement for housing. It was assumed that these measures would drive poverty entrepreneurs out of business, and that the phenomenon would be eventually eradicated. But in fact it was the lodging households that were affected the most. The first measures on "no-benefit zones" were implemented in 2017. While cities generally have approved and praised the measure, it in fact led to a more precarious situation for lodgers (Žaneta Motlová, 2018; Česká televize, 2019).

A similar attempt to curb poverty entrepreneurship was made in 2019 when stricter rules on the allocation of both the supplement for housing and the housing allowance were introduced. The immediate effect was to limit access to both welfare mechanisms (Monika Veselíková, 2018). This measure was highly criticized by the beneficiaries in interviews with the authors. However, at the time of writing, it is not possible to gather more comprehensive data on the effect of the policy.

Though these are the only instruments provided by the state to solve the housing issue for the most vulnerable, there are many examples of good practice in social housing initiated by municipalities. Some municipalities complemented the blanket solution of "no-benefit zones" with long-term social assistance and provision of social housing. The city of Brno relocated 50 families to standard rent apartments owned by the municipality. In addition to the city of Brno, the project involved social workers, NGOs, and the tenants themselves, collectively finding solutions so that families could keep their houses for one year (Černá *et al.*, 2018). In 2017, after 1 year of project implementation, 48 out of 50 families were able to keep their dwelling. Research performed over the implementation period revealed several other positive benefits, such as reduced health problems and more personal security. The introduction of energy metering and consultancy on how to benefit from the information provided by it allowed tenants to keep better track of their own energy expenses (Ripka *et al.*, 2018). A similar approach is currently under preparation in 16 other Czech municipalities, ranging from big cities to small villages. But as local issues and capabilities vary, implementation is also diverse. Implementation in the city of Ostrava, for example, involves more stakeholders, including energy providers. In some cases where tenants are indebted to energy providers (especially heat providers), the city, with the support of social workers, negotiates debt repayment and provides guarantees to the energy company in order to allow households to be reconnected to energy (Ministry of Labour and Social Affairs, 2019c).

Energy efficiency measures

In the Czech Republic energy efficiency is loosely related to energy poverty. However, for energy poor households, an increase in energy efficiency provides either an opportunity to reduce or abandon self-restriction on energy consumption or to provide savings in energy expenditure. These benefits of energy efficiency are well known and acknowledged (International Energy Agency, 2014). Looking at the overall indicators, the country does not belong to the best performers in energy efficiency – its energy intensity is the fourth highest in the EU (Statistical Office of the European Communities, 2019). The Czech Republic has the fifth highest overall energy consumption of energy per dwelling in the EU (Odyssee-Mure, 2019). However, there are a large number of programs aimed at improving energy efficiency and at increasing energy savings available (Karásek and Pojar, 2018). It is important to understand how these programs target energy poor households.

Under EU regulations, most notably under the so-called Energy Efficiency Directive 2012/27/EU (EED), the Czech Republic is compelled to implement certain energy efficiency measures in order to reach the target of decreasing energy consumption by 1.5 percent per year. Two instruments are designed to help meet these obligations. The first such instrument, the "energy efficiency obligation scheme," puts emphasis on market measures, where energy service providers are mandated to decrease energy consumption. The second instrument is using "alternative measures," through which state governing bodies are tasked with achievement of energy savings instead (Directive 2012/27/EU). In order to prevent increases in end-user energy prices, the Czech Republic chose to implement alternative measures to fulfill its obligations (Ministry of Industry and Trade, 2014). The government designed a number of energy savings policies to be implemented using European Cohesion Funds. Decision-makers made good use of the overlap between the implementation period of the Energy Efficiency Directive and the Cohesion Funds budget, 2014 to 2020 (Borshchevska and Lehotský, 2015). The Ministry of Industry and Trade was put in charge of implementation. It designed a list of strategies, action plans, and programs, such as the State Energy Concept, or the Energy Efficiency National Action Plan (EENAP). With their 2016 and 2017 updates, these documents became the fundaments of action. The indicative target established by the 2016 EENAP was 50.67 PJ of energy savings by 2020, and annual increases of savings afterwards (Ministry of Industry and Trade, 2016a). Most of these savings are to be pursued through investment activities. Individuals or organizations are incentivized to invest in refurbishment or retrofitting by using essentially two types of financing programs. The first type is based on the use of EU Structural Funds (operational programs, henceforth OPs), and the second type on national resources. These are allocated through financial grants (Sochor, 2017). All relevant efficiency-related programs are summarized in Table 9.1.

From an energy poverty perspective, the decision of the Czech government to steer away from market-based approaches to achieve energy savings was very promising. The expected short-term adverse impact on energy prices, often cited as a main reason to avoid taxes on inefficient consumption, was prevented. Moreover, some alternative programs were already in place and delivering well (Ministry of Industry and Trade, 2014). The state was in a position to design funding programs with which to target the people most in need. However, the potential of alleviating energy poverty has not been fully unlocked so far.

Table 9.1 shows that EU funds were expected to deliver most of the energy savings. However, Czech OPs were approved long after their original schedule and after the start of the implementation period of the

Table 9.1 Overview of efficiency-related programs

Program name	*Time frame*	*Directing body*	*Financing source*[a]	*Financial allocation planned (EUR)*[b]	*Financial allocation actual (EUR)*[b]	*Savings – NAPEE proposal (PJ)*	*Savings – actual prediction (PJ)*
Enterprises and Innovations OP	2015–2023	Ministry of Industry and Trade	ERDF	779,000,000	674,000,000	20.0	7.6
Environment OP	2015–2023	Ministry of Environment	ERDF, CF	935,000,000	919,000,000	5.0	5.0
Integrated Regional OP	2016–2023	Ministry of Regional Development	ERDF	623,000,000	374,000,000	2.0	3.5
New Green Savings	2014–2021	Ministry of Environment	Emission allowances	1,052,000,000	896,000,000	14.3	10.3
EFEKT Programme	2017–2021	Ministry of Industry and Trade	State budget	27,000,000	31,000,000	0.4	1.6

Notes

a ERDF – European Regional Development Fund; cf. – Cohesion Fund.

b Average 2019 EUR–CZK conversion rate was used (1 EUR = 25,672 CZK).

Energy Efficiency Directive. Both "The Enterprises and Innovations Operational Program" and "The Environment Operational Program" that were supposed to carry through these objectives were among the last Ops to be approved by the government and only received a green light from Brussels at the beginning of 2015 (Operační program Podnikání a inovace pro konkurenceschopnost 2015b). Due to a slow completion process and weak leadership and cross-coordination (Shabu, 2015) these programs delivered in a suboptimal way. In addition, calls for projects were delayed even further (Sochor, 2015). Given the situation, implementation of these programs is, in the end, expected to favor those projects which are faster to implement and which will provide clearer energy efficiency gains at the cost of supporting projects that might aid vulnerable consumers more.

Energy efficiency implementation is also largely affected by the way programs are designed. Implementation was and still is the responsibility of multiple agencies. The main body responsible for energy efficiency issues is the Ministry of Industry and Trade. It oversees implementation of the "Enterprise and Innovations Operational Program" focused on industrial players and of the national "EFEKT" program focused on non-investment projects. The Ministry of Environment administers the "Environment Operational Program" and the "Green Savings" scheme, both aiming for lower air pollution through energy efficiency investments – the first one in both public buildings and residential houses, the second in residential houses only. The Ministry of Regional Development oversees the "IROP" scheme, which finances energy efficiency intervention in residential condominium houses and is responsible for implementation of policies in the housing sector at large (Borshchevska and Lehotský, 2015). Given this institutional ecosystem and diversity of programs with sometimes overlapping responsibilities, it is difficult to successfully follow through on action.

As they are designed, most of these programs are difficult for potential beneficiaries to understand and access. Most are based on financial grants, which require co-financing. Moreover, a potential grantee has to fulfill a list of conditions in order to become eligible for a grant. If the application is successful, the grantee undertakes investment measures (e.g., the installation of boilers or outer shell retrofitting) from their own resources and is reimbursed only after investment completion. The Green Savings scheme and the Environment Operational Program for individual households are predominantly grant-based. The only exceptions are pilot projects and low-interest loans to house owners offered in smaller schemes (e.g., PANEL 2013+). Despite the quite short return on investment of partial retrofitting projects, paying for grants in advance is not always an option, especially for the more vulnerable categories identified before.

Setting aside the requirement for grantees to use their own resources, the access to grants also depends, to a large extent, on individual ability to navigate through programs. The boiler replacement program illustrates this well. To receive a grant for boiler replacement, the applicant needs to access the Green

Savings scheme administered by the State Environmental Fund under the Ministry of Environment. One requirement is to submit project documentation on expected energy efficiency gains. Another is to dispose of investment funds in advance (State Environmental Fund, 2019). There is also funding available for project documentation commissioning under the "EFEKT" program administered by the Ministry of Industry and Trade (2019). In addition, there is a third "Boiler Subsidy" program targeting old polluting solid fuel boilers, administered by the Ministry of Environment, where there's no requirement to submit project documentation. Similarly, subsidies for apartment building retrofits are offered through the Green Savings program and on top of that the Ministry of Regional Development offers low-interest loans for the same purpose. Overlapping programs vary slightly in eligibility conditions, types, and levels of support and other details (Borshchevska and Lehotský, 2015). But property owners need to allocate time to learn the requirements and differences among these programs or find external expertise in order to submit successful applications. Such options are not within the reach of people struggling with energy poverty and income poverty.

There is, however, one exception – the Boiler Subsidy program – which targets the most energy vulnerable consumers. In the standard program scenario, property owners cover the investment using their own resources, while being partially reimbursed for up to 80 percent of the cost, but only after the investment has been finished. However, as mentioned, the upfront cost of investment prevents many from accessing the subsidy. For that reason, three Czech regions (Usti, Karlovy Vary, and Moravia Silesia) are, at the moment of this writing, testing an alternative scheme of a "boiler loan" paired with a "boiler subsidy" based on which municipalities identify potential beneficiaries, who are then assisted in their loan application. This loan helps cover all investment costs. The reimbursement is given in installments to coincide with the loan repayment agenda (State Environmental Fund, 2019). The project is in its very beginnings, and thus there is no data on its success or any external evaluation available.

To conclude, there are three typical features of energy poverty in the Czech Republic. First, conventional indicators do not accurately display the existence of the phenomenon. More careful observers can conclude that energy poverty affects households rather individually. Given their low income and socio-demographic characteristics, affected households struggle to meet their energy and other household needs. Second, energy poverty in the country is related to the technical characteristics of the building stock. The lowest income households living in old uninsulated single-family houses are expected to be the most vulnerable. Last, the geographical occurrence of energy poverty is uneven. Energy poverty can be identified in particular areas and locations where socio-economic transformations accumulated, over time, to create multiple forms of social disadvantages.

In this chapter, we claim that due to the "hidden" nature of energy poverty, the risk prevails that specific strata of the population that in fact experience energy deprivation to a large extent can be misrecognized in state policies, as summarized above. Given the situation, evidence shows that two types of policies are reasonably capable of addressing the issue. However, so far, measures have not lived up to their full potential.

The first type of policy is related to the housing needs of the population. On the one hand, the housing allowance subsidy, which covers expenses on rent and/or energy, prevents energy poverty from manifesting in households with a low income but not directly threatened by poverty. On the other hand, there is also a significant part of the population living in dire conditions in either so-called socially excluded localities or in shelters. Their marginalization is caused either by the lack of social housing or by high debts (including energy debts), which in the end may exclude them from housing market. The lack of social housing and the absence of rehabilitation mechanisms for severely indebted people creates an opening for the emergence of so-called poverty entrepreneurship, under which individuals are subject to illicit housing practices. Moreover, attempts to limit this phenomenon in the past resulted in the opposite effect, increasing the vulnerability of this segment.

The second type of policy addresses the energy efficiency of the housing stock. While state subsidies are useful for the majority of the population, they are largely inaccessible for energy poor households. Access to grants depends to a large extent on an individual's ability to navigate through them. What is more, the majority of programs are based on financial grants, which require of the households both co-financing and initial investment funds. For the more vulnerable categories identified in this chapter, paying for efficiency measures in advance is, most of the time, not an option due to lack of savings.

Therefore, even though energy poverty is not severe and affects comparatively few households, the policies in place do not mitigate the phenomenon, but rather facilitate the growth of inequalities and reinforce the structural nature of energy poverty in the Czech Republic. There are, however, some good practices at the national as well as regional level, which aim to target vulnerable households better, such as the pilot schemes for boiler replacement; initiatives of local actors who provide accessible housing together with long-term social support; or initiatives of cities who act as intermediaries between energy providers and debt-prone tenants.

Notes

1 According to the 2018 survey, housing costs as a proportion of households' net money income reached 15.9 percent on average, down from 18.3 percent in 2014 (Czech Statistical Office, 2019a). Energy expenses as a percentage of households' net money income reached 9.3 percent on average in 2018 (authors' own calculation based on CSO statistical data).

2 In 2018, 1.5 percent of Czech households declared having arrears on paying rent and housing costs of apartment owners (if recounted from 59.2 percent of applicable households, it would be approx. 2.5 percent); 1.8 percent acknowledged having

arrears on paying heating, electricity, gas, and water (from 99.8 percent of applicable households); 0.3 percent had issues with repaying mortgages and other dwelling-related loans (if recounted from 21.1 percent of applicable households, it would be approx. 1.4 percent); and 0.7 percent declared having arrears on repaying other loans (if recounted from 11.7 percent of applicable households, it would be approx. 6 percent) (Czech Statistical Office, 2019e).

3 Enforcement proceedings (EP) represent a serious issue for more and more Czech households and entrepreneurs. In 2017, there were 9.7 percent or 863,000 natural persons in EP proceedings in the Czech Republic. Compared to the year 2016, the number of people increased by 3.4 percent. Debt cycling is understood as one of the major causes. Another issue is multiple EPs. In 2017, there were 493,000 natural persons with 3 or more EPs in the Czech Republic (Ecumenical Academy Prague, 2019).

4 Most of the family houses are located in smaller municipalities in rural areas. Although on average the total housing expenditures are lower in smaller municipalities than in large cities, this is caused by the fact that the total housing costs in the cities are substantially increased by the cost of rent. If we look only at energy expenditures, they are higher in smaller municipalities than in large cities (Karásek *et al.*, 2016, pp. 37–38; Czech Statistical Office, 2019a).

5 Against 14.6 percent of lowest-income households living in family houses with wall insulation; 11.6 percent occupying family houses with roof insulation; and 36.3 percent of households living in family houses with thermally insulated windows (Karásek *et al.*, 2016, pp. 37–38; Czech Statistical Office, 2019a).

6 Against 28.9 percent of lowest-income households living in apartments in apartment buildings with wall insulation; 17.9 percent inhabiting apartment buildings with roof insulation; and 44.6 percent of households living in apartment buildings with thermally insulated windows (Czech Statistical Office, 2017, p. 94).

7 In the Czech Republic, family houses (usually in private ownership) are the most prevalent in smaller municipalities (chart in Karásek *et al.*, 2016, p. 46).

8 Based on the differences in the issue of social disadvantage in urban and rural areas, only rural areas are part of the research here (Bernard and Šimon, 2017, p. 9).

9 The highest proportion of the population over 15 years under EP can be found in the Ústecký (18.1 percent) and Karlovarský (17.4 percent) regions. On the other hand, in Zlín (5.7 percent) and Vysočina (5.8 percent) regions, people are least affected (Ecumenical Academy Prague, 2019).

References

Bernard, J. and Šimon, M., 2017. Inner peripheries in the Czech Republic: the multidimensional nature of social exclusion in rural areas. *Czech Sociological Review*, 53 (1), 3.

Borshchevska, Y. and Lehotský, L., 2015. Evaluating the Czech Republic's progress on energy efficiency: major achievements and pitfalls. *International Issues and Slovak Foreign Policy Affairs*, Vol. XXIV (No. 4).

Bouzarovski, S. and Tirado Herrero, S., 2017. Geographies of injustice: the socio-spatial determinants of energy poverty in Poland, the Czech Republic and Hungary. *Post-Communist Economies*, 29 (1), 27–50.

Černá, E., Ripka, Š., and Bartošová, I., 2018. *Pilotní testování rychlého zabydlení rodin s dětmi (Rapid Re-Housing): Analýza výchozího stavu účastníků z cílové skupiny projektu.* Ostrava.

Červenka, J. and Ďurďovič, M., 2018. Veřejnost o spotřebě a úsporách energie, tepla a vody ve svých domácnostech – květen 2018 [online]. Available from: https://cvvm.soc.cas.cz/media/com_form2content/documents/c2/a4719/f9/oe180913c.pdf [Accessed 18 Oct 2019].

Česká televize, 2019. Města si pochvalují bezdoplatkové zóny. Tím se ale problém neřeší, varuje Institut pro sociální inkluzi [online]. *ČT24*. Available from: https://ct24.ceskatelevize.cz/regiony/ustecky-kraj/2871956-mesta-si-pochvaluji-bezdoplatkove-zony-tim-se-ale-problem-neresi-varuje [Accessed 13 Oct 2019].

Czech Statistical Office, 2017. Spotřeba paliv a energií v domácnostech [online]. Available from: www.czso.cz/documents/10180/50619982/ENERGO_2015.pdf [Accessed 17 Oct 2019].

Czech Statistical Office, 2019. Living Standards, Household Consumption [online]. Available from: www.czso.cz/csu/czso/living_conditions_household_income_and_expenditure_stat [Accessed 18 Oct 2019].

Czech Statistical Office, 2019a. Životní podmínky 2018 [online]. Available from: www.czso.cz/documents/10180/111272138/csu_tk_silc_press_kit.pdf/37e33370-b9aa-4ace-a272-2edf6d88da39?version=1.0 [Accessed 18 Oct 2019].

Czech Statistical Office, 2019b. Životní podmínky 2018: Domácnosti podle decilového rozdělení čistých peněžních příjmů na osobu – Charakteristiky bydlení a vybavenost domácnosti (%) [online]. Available from: www.czso.cz/documents/10180/111272138/csu_tk_silc_press_kit.pdf [Accessed 17 Oct 2019].

Czech Statistical Office, 2019c. Životní podmínky 2018: Domácnosti podle postavení osoby v čele – Charakteristiky bydlení a vybavenost domácnosti (%) [online]. Available from: www.czso.cz/documents/10180/111272138/csu_tk_silc_press_kit.pdf.

Czech Statistical Office, 2019d. Životní podmínky 2018: Domácnosti podle druhu domácnosti EU – Charakteristiky bydlení a vybavenost domácnosti (%) [online]. Available from: www.czso.cz/documents/10180/111272138/csu_tk_silc_press_kit.pdf.

Czech Statistical Office, 2019e. Životní podmínky 2018: Domácnosti podle decilového rozdělení čistých peněžních příjmů na osobu – Subjektivní názory (%) [online]. Available from: www.czso.cz/documents/10180/91839453/160021192e.pdf.

Czech Statistical Office, 2019f. Životní podmínky 2018: Domácnosti podle postavení osoby v čele – Subjektivní názory (%) [online]. Available from: www.czso.cz/documents/10180/91839453/160021191e.pdf.

Czech Statistical Office, 2019g. Životní podmínky 2018: Domácnosti podle druhu domácnosti EU – Subjektivní názory (%) [online]. Available from: www.czso.cz/documents/10180/91839453/160021199e.pdf.

Czech Statistical Office, 2019h. Životní podmínky 2018: Domácnosti podle krajů – 1. část – Subjektivní názory (%) [online]. Available from: www.czso.cz/documents/10180/91839453/16002119141e.pdf.

Czech Statistical Office, 2019i. Životní podmínky 2018: Domácnosti podle krajů – 2. část – Subjektivní názory (%) [online]. Available from: www.czso.cz/documents/10180/91839453/16002119142e.pdf.

Directive 2012/27/EU of the European Parliament and of the Council of 25 October 2012 on energy efficiency, amending Directives 2009/125/EC and 2010/30/EU and repealing Directives 2004/8/EC and 2006/32/EC Text with EEA relevance [online], 2012. Available from: http://data.europa.eu/eli/dir/2012/27/oj [Accessed 10 Sep 2016].

Ecumenical Academy Prague, 2019. Mapa Exekucí [online]. Available from: http://mapaexekuci.cz/index.php/map-en/ [Accessed 17 Oct 2019].

European Commission, 2019. Czech Republic – Housing allowance [online]. Available from: https://ec.europa.eu/social/main.jsp?catId=1106&langId=en&intPageId=4470 [Accessed 12 Oct 2019].

Filčák, R. and Živčič, L., 2017. Energy poverty and multi-dimensional perspectives of social inequalities and policy challenges. *Slovak Foreign Policy Affairs*, 26 (1–2), 40–61.

Foldynová, I., Hruška, L., Hrušková, A., Krhutová, M., Fujak, R., Ševčík, J., and Dědič, L., 2015. *Podkladová analýza ke koncepci bydlení statutárního města Ostravy*. Ostrava.

Gabal, I. and Víšek, P., 2010. *Východiska strategie boje proti sociálnímu vyloučení: Východiska pro formulaci a implementaci politiky začleňování obyvatel vyloučených lokalit do české společnosti a její sociální a ekonomické struktury*. Praha.

Government of the Czech Republic, 2017. Vláda schválila návrh zákona o sociálním bydlení [online]. Available from: www.vlada.cz/cz/clenove-vlady/pri-uradu-vlady/jan_chvojka/aktualne/vlada-schvalila-navrh-zakona-o-socialnim-bydleni-154194/ [Accessed 11 Oct 2019].

Hampl, M., 2007. Regionální diferenciace současného socioekonomického vývoje v České republice/Regional Differentiation of Current Socio-Economic Development in the Czech Republic. *Sociologický Časopis/Czech Sociological Review*, 43 (5), 889–910.

Hampl, M. and Marada, M., 2015. Sociogeografická regionalizace Česka. *Geografie*, 120 (3), 397–421.

Hruška, L., Foldynová, I., Hrušková, A., Doleželová, H., Fujak, R., Proske, P., Drijáková, I., Baranová, M., and Štecher, R., 2013. *Analýza a řešení problematiky ubytoven na území statutárního města Ostravy v krátkodobém a dlouhodobém horizontu*. Ostrava.

Hruška, V., 2014. Diferenciace venkovského prostoru na příkladu Moravskoslezského kraje. Masarykova univerzita, Přírodovědecká fakulta, Brno.

International Energy Agency, 2014. *Capturing the Multiple Benefits of Energy Efficiency*. 2nd ed. Paris: IEA.

Karásek, J., Krivošík, J., Pojar, J., and Anisimova, N., 2016. Opatření proti energetické chudobě v ČR [online]. Available from: www.mpo-efekt.cz/upload/7799f3fd595eeee1fa66875530f33e8a/energeticka-chudoba-v12.pdf [Accessed 17 Oct 2019].

Karásek, J. and Pojar, J., 2018. Programme to reduce energy poverty in the Czech Republic. *Energy Policy*, 115, 131–137.

Mikeszová, M., Lux, M., and Morisseau, A., 2009. Potenciální finanční nedostupnost nájemního bydlení po deregulaci nájemného – regionální perspektiva. *Czech Sociological Review*, 45 (2), 315–343.

Ministry of Industry and Trade, 2014. Národní akční plán energetické účinnosti ČR: dle čl. 24 odst. 2 směrnice Evropského parlamentu a Rady 2012/27/EU ze dne 25. října 2012 o energetické účinnosti [online]. Available from: http://download.mpo.cz/get/50711/63238/651839/priloha003.pdf [Accessed 27 Sep 2016].

Ministry of Industry and Trade, 2016a. Národní akční plán energetické účinnosti ČR 2016 [online]. Available from: www.mpo.cz/assets/dokumenty/50711/63238/651838/priloha004.pdf [Accessed 6 Sep 2016].

Ministry of Industry and Trade, 2016b. *Státní program na podporu úspor energie na období 2017–2021 Dotační titul: PROGRAM EFEKT*. Prague.

Ministry of Industry and Trade, 2019. *Státní program na podporu úspor energie na období 2017–2021*. Praha.

Ministry of Labour and Social Affairs, 2019a. Assistance in Material Need [online]. Available from: www.mpsv.cz/web/en/assistance-in-material-need [Accessed 12 Oct 2019].

Ministry of Labour and Social Affairs, 2019b. State Social Support [online]. Available from: www.mpsv.cz/web/en/state-social-support [Accessed 17 Oct 2019].

Ministry of Labour and Social Affairs, 2019c. Dobrá praxe v ČR: Síťování a rozvoj odborných kompetencí obcí v sociálním bydlení [online]. Available from: www.socialnibydleni.mpsv.cz/cs/co-je-socialni-bydleni/priklady-dobre-praxe/2-uncategorised/103-pracovni-fokusni-expertni-skupiny [Accessed 12 Oct 2019].

Ministry of Regional Development, 2018. *Bydlení v České republice v číslech (srpen 2018)*. Prague: Ministry of Regional Development.

Monika Veselíková, 2018. Příspěvek i doplatek přísněji. Změny v dávkách na bydlení [online]. *Peníze.cz*. Available from: www.penize.cz/prispevek-na-bydleni/336569-prispevek-i-doplatek-prisneji-zmeny-v-davkach-na-bydleni [Accessed 12 Oct 2019].

Musil, J. and Müller, J., 2008. Vnitřní periferie v České republice jako mechanismus sociální exkluze/Inner Peripheries of the Czech Republic as a Mechanism of Social Exclusion. *Sociologický Časopis/Czech Sociological Review*, 44 (2), 321–348.

Novák, J. and Netrdová, P., 2011. Prostorové vzorce sociálně-ekonomické diferenciace obcí v České republice. *Sociologický Časopis/Czech Sociological Review*, 47 (4), 717–744.

Odyssee-Mure, 2019. Consumption per dwelling scaled to EU average climate: 2016 [online]. Available from: www.indicators.odyssee-mure.eu/energy-indicators/climate-consumption-dwelling.html [Accessed 17 Oct 2019].

Operační program Podnikání a inovace pro konkurenceschopnost, 2015a. Indikativní harmonogram výzev na rok 2016 – Operační program Podnikání a inovace pro konkurenceschopnost [online]. Available from: www.oppik.cz/files/609-op-pik-harmonogram-vyzev-2016-5232-1.pdf [Accessed 25 Apr 2017].

Operační program Podnikání a inovace pro konkurenceschopnost, 2015b. OPPIK má od Evropské komise zelenou [online]. Available from: www.oppik.cz/aktuality/oppik-ma-od-evropske-komise-zelenou [Accessed 15 Sep 2016].

Prokop, D., 2019. Does Housing Matter? Inadequate Housing and School Performance of Children from Poor Households in the Czech Republic. *Czech Sociological Review*, 55 (4), 445.

Ripka, Š., Černá, E., and Kubala, P., 2018. *Pilotní testování rychlého zabydlení rodin s dětmi (Rapid Re-Housing): Závěrečná evaluační zpráva/Analýza cílového stavu. Dopady zabydlení po 12 měsících od nastěhování*. Ostrava.

Schvarczová, H., 2019. Nástroje státu na snižování spotřeby energie [online]. *Financing Energy Efficiency in Central Europe*. Available from: www.egubrno.cz/wp-content/uploads/2019/11/seminar_EFEKT2019_sbornik_web.pdf [Accessed 22 Feb 2020].

Shabu, M., 2015. Spřípravou na čerpání peněz z EU fondů jsme na tom špatně [online]. *Česká pozice*. Available from: http://ceskapozice.lidovky.cz/s-pripravou-na-cerpani-penez-z-eu-fondu-jsme-na-tom-spatne-pz9-/tema.aspx?c=A150507_153141_pozice-tema_lube [Accessed 25 May 2017].

Sochor, V., 2015. Strategie ČR v plnění národních cílů Směrnice o energetické účinnosti – stávající a budoucí nástroje podpory [online]. Available from: www.svn.cz/assets/files/seminare_a_konference/2015/listopad-2015/2_VSochor.pdf [Accessed 27 Sep 2016].

Sochor, V., 2017. State of play of energy efficiency investment and financing scheme – Czech Republic [online]. *Financing Energy Efficiency in Central Europe*.

State Environmental Fund, 2019. About the New Green Savings Programme [online]. Available from: www.novazelenausporam.cz/about-the-new-green-savings-programme/ [Accessed 17 Oct 2019].

State Environmental Fund, 2019. *Výzva č. 1/2019 k předkládání žádostí o poskytnutí podpory v rámci programu na podporu výměny nevyhovujících kotlů na pevná paliva v domácnostech, úspor energie a dalších adaptačních či mitigačních opatření ve vztahu ke změně klimatu v obcích Karlovarskéh*. Praha.

State Housing Development Fund, 2017. Program Panel 2013+ [online]. Available from: www.sfrb.cz/programy-a-podpory/program-panel-2013/ [Accessed 23 May 2017].

Statistical Office of the European Communities, 2019. Energy intensity: nrg_ind_ei [online].

Sunega, P. and Lux, M., 2018. Equal in Incomes, Unequal in Wealth? Housing Wealth Inequalities in the Czech Republic. *Czech Sociological Review*, 54 (5), 749.

Szénássy, E., Kňapová, K., and Dvořáková, T., 2017. *Jak zůstat člověkem v síti dluhů: Kvalitativní výzkum*. Praha.

Thomson, H. and Bouzarovski, S., 2018. Addressing Energy Poverty in the European Union: State of Play and Action [online]. Available from: www.energypoverty.eu/sites/default/files/downloads/publications/19-05/paneureport2018_updated2019.pdf [Accessed 17 Oct 2019].

Valentová, B., 2005. Vývoj sociálního bydlení s důrazem na jeho zacílení na určité sociální vrstvy obyvatelstva/The Development of Social Housing with a Focus on Targeted Housing for Certain Social Strata. *Sociologický Časopis/Czech Sociological Review*, 41 (2), 301–315.

Zákon č. 111/2006 Sb.: Zákon o pomoci v hmotné nouzi, 2019.

Žaneta Motlová, 2018. Bezdoplatkových zón v kraji přibývá. Města už vidí první výsledky [online]. *iDNES.cz*. Available from: www.idnes.cz/ostrava/zpravy/bezdoplatkova-zona-ubytovna-chudoba.A180327_391998_ostrava-zpravy_woj [Accessed 13 Oct 2019].

10 Divided we (still) stand?

Energy poverty in East and West Germany

Philipp Biermann

Introduction

German culture and society are historically affected by two world wars resulting in a temporary division in occupied zones after World War II. This division was manifested not only in terms of geography but also in terms of culture and ideology. The western part was occupied by the capitalist democratic countries France, the United States (US), and the United Kingdom (UK) resulting in the Federal Republic of Germany (FRG), while the eastern part was occupied by the communist Soviet Union (USSR) resulting in the German Democratic Republic (GDR). A body of literature states that these differences persist even after almost 30 years of reunification in 1990.[1]

Although there are no institutional differences anymore in terms of governance and government structure, the division of east and west is showing up in some economic indicators, but even more at the individual level. The literature distinguishes differences in terms of circumstances and mentality. While the former is linked to economic and socio-demographic "measurable" indicators, the latter is studied by analyzing data at the individual level regarding psychological and other subjective indicators of human well-being (e.g. Welsch and Biermann, 2019). The differences in mentality in particular are usually connected to the era of division of Germany into FRG (West) and GDR (East).

Previous studies have found differences in the socio-economic conditions and in the mentality of East and West Germans. Some of those issues are rooted in the communist heritage of East Germany, while others are related to people's feeling that they are (still) left behind the West in the aftermath of reunification. The difference in the circumstances shows up in the economic and demographic indicators in East and West Germany and is seen as a persistence of disadvantages from the socialist era, although there is a catching-up process. The feeling of being left behind is related to people from the East having a different mentality in terms of their position in the reunified society. Large investment and institutional efforts had to be undertaken to reunify the two economically and ideologically different systems. It requested also significant monetary transfers from West to East after 1990. This created a situation where West Germany was regarded as a donor while East Germany was regarded as a receiver. It manifested in a mentality difference in people living in East and West Germany. In some surveys we can find these differences still today after 30 years.

The interesting question in the light of those findings is whether this circumstance and mentality difference would also show up regarding energy poverty. To shed some light on this issue, this chapter is meant to distinguish energy poverty into circumstances and mentality. Different circumstances of energy poverty would show up in differences in the energy poverty rates while mentality is to be represented by the effect energy poverty has on individual well-being (Biermann, 2016). The two questions to be analyzed in this chapter are:

1 Is the rate of energy poverty incidence different in East and West Germany?
2 Does the effect of energy poverty on subjective well-being differ between East and West?

While differences in the energy poverty rates between East and West represent differences in circumstances, a difference in the well-being effect would, *ceteris paribus*, imply a difference in mentality regarding energy poverty. The distinction between circumstance and mentality effects regarding subjective well-being in the East and West was, for example, used by Biermann and Welsch (2019).

Data and method

To quantify energy poverty and subjective well-being, we use the German Socio-Economic Panel Dataset (SOEP), a large survey hosted at DIW in Berlin (Wagner et al., 2007). This survey has the advantage of having a panel structure interviewing about twenty thousand individuals repeatedly every year. The dataset embodies a large set of indicators on socio-demographic, economic, and psychological information at both the individual and the household level. Furthermore, it contains information on household expenditure on energy services (e.g., heat and warm water, electricity). With this information it is possible to get a picture both of differences in energy poverty rates and in the well-being effect of individual energy poverty incidence (Biermann, 2016).

Using the SOEP, Heindl (2015) compares energy poverty rates in accordance with the type of measurement.[2] He finds that the share of the population in energy poverty differs by measurement type. Therefore, we are considering different types of measurement for our analysis. Biermann (2016) clustered energy poverty measures into different "types." There are measures based on high absolute energy expenditure (type 1), measures based on relative energy expenditures (type 2), and the low-income high-cost

(LIHC, type 3) indicator. Type 1 and type 2 measures are then further distinguished in their reference point of energy expenditure being the mean or the median expenditure of the respective population. LIHC is a measure that needs two conditions to be fulfilled for a household to be regarded as energy poor. The household must exceed a threshold of energy expenditure and at the same time be affected by income poverty (Hills, 2012).

The econometric model we have chosen for our analysis is a panel based within regression, where we control for fixed effects at the individual level. Effects that are fixed within a person over time regarding her well-being are controlled for by measuring the effect of the derivation of any independent variable from its idiosyncratic mean over time.[3] To explain the method short and without being exhaustive, we can say that we have excluded the possibility that the effects we estimate in our model are driven by person-related idiosyncratic differences. The advantage of such a procedure is avoiding interpreting statistical artifacts resulting from interpersonal differences, therefore making individuals more comparable regarding the well-being effects of the independent (explanatory) variables. These models are not always used in survey analysis, because they have rather high requirements on the data. Fortunately, the SOEP gives us the opportunity to apply the aforementioned method due to its panel structure at the individual level (Wagner et al., 2007). The life satisfaction equation that we estimate is as follows:

$$LS_{it} = \alpha_i + \gamma_t + \beta^* fuelpoverty_{it} + \delta' \mathbf{controls}_{it\,t} + \lambda^* netinc_{it} + \omega^* incpoverty_{it} + \varepsilon_{it},$$

where LS_{it} is the reported life satisfaction of individual i at time t. The vector $\mathbf{controls}_{it}$ denotes individual observed characteristics that vary over time.[4] Our $fuelpoverty_{it}$ dummy denotes whether the household the individual lives in was affected by fuel poverty at the time of the interview (incidence of fuel poverty). The variables $netinc_{it}$ and $incpoverty_{it}$ capture equivalized net monthly household income and income poverty respectively. While α_i denotes the individual fixed effects, γ_t captures the time fixed effects and ε_{it} denotes the individual and time specific error term. This model is in line with the general approach to analyzing subjective well-being empirically (see Ferrer-i-Carbonell and Frijters (2004) for a discussion of methodology with respect to subjective well-being analysis).

Empirical findings

Following Biermann (2016), there are four different measures of energy poverty which can be assigned to the three types described above. Following their definitions described by Heindl (2015), the two-times-median-expenditure measure (2*median X) defines an individual as energy poor when her household's expenditure for energy services exceeds the double the median expenditure of the reference group (which is in our case the German population). The 10 percent share of income measure (10%) defines an individual as energy poor if her household's expenditure for energy services exceeds ten percent of its income, whereas the two-times-median-expenditure-share measure (2*median XS) defines an individual as affected by energy poverty when her household's expenditure for energy services in relative terms (as share of income) exceeds twice the median of the energy expenditure share of the reference population. Finally, the LIHC measure defines an individual as energy poor if both the energy expenditure share of the household exceeds the median and the household is regarded as income poor at the same time. 2*mean X represents type 1 measures, 10% and 2*median XS represent type 2 measures, and LIHC represents type 3 measures.[5] The respective energy poverty rates of Germany overall and distinguished by East and West are shown in Table 10.1.

This table represents the circumstances of individuals regarding energy poverty. The results show that energy poverty rates are higher in East Germany than in West Germany for type 2 and type 3 measures. The share of households affected by the type 1 measure of energy poverty is lower in East than in West Germany. This is due to its absolute-expenditure-based nature. The main critique about this approach is that it does not take into account any criteria other than absolute expenditures on energy. It is obvious that this kind of measurement is not effective in representing marginalized groups, since it would also count households as energy poor if they spend a lot on energy services due to individual preferences or simply because they have a very high income and do not care about the level of

Table 10.1 Energy Poverty Rates (EPR) by measure

Measure	*EPR total (%)*	*EPR East (%)*	*EPR West (%)*
2*median X	7.69	6.21	8.16
10%	20.06	23.01	19.11
2*median XS	11.10	13.31	10.39
LIHC	6.33	7.90	5.82

Note
Data from Goebel et al., 2019.

Table 10.2 Economic situation in East and West

Measure	*EPR total*	*EPR East*	*EPR West*
Unemployment	5.76%	9.36%	4.61%
Net income (mean)	€2,712	€2,216	€2,870
Income poverty	10.89%	14.89%	9.60%

Note
Data from Goebel et al., 2019.

their expenditure on energy. The prevalence of higher energy poverty (type 2 and type 3) is in line with the overall findings on other economic circumstances, being worse in East than in West Germany (see for example Biermann and Welsch, 2019). This is also confirmed in the present case as shown in Table 10.2. East Germany has a higher prevalence of income poverty and unemployment, and a lower average income.

This is of high relevance when we look at the well-being effect of those circumstances in the following empirical analysis. Since the type 1 measure does not represent marginalized households, we do not take it into account in our empirical analysis. Furthermore, we only investigate one of the type 2 measures, the 2*median XS measure, since it is our preferred measure from this group.[6] Against this background, the second research question is analyzed with the help of a multivariate regression analysis following the one conducted in Biermann (2016). We run a regression as in the equation shown above to estimate the correlation between life satisfaction and energy poverty, controlling for several socio-economic factors. The results are presented in Tables 10.3a and 10.3b, where Table 10.3a shows the effect of energy poverty on well-being. In Table 10.3b we additionally control for income poverty, which means that the effect (coefficient) of energy poverty is to be interpreted as independent from the income deprivation effect, where people are affected by income and energy poverty at the same time. Due to the *ceteris paribus* assumption of those models, we can say that this is the income-independent effect of energy poverty. Further, it gives us the

Table 10.3a Differences in mentality regarding energy poverty

	A (East)	*B (West)*	*C (East)*	*D (West)*
Poor health	–0.467***	–0.497***	–0.468***	–0.497***
	(0.00920)	(0.00520)	(0.00920)	(0.00520)
Unemployed	–0.516***	–0.504***	–0.516***	–0.502***
	(0.0241)	(0.0203)	(0.0241)	(0.0203)
Equivalized income	0.000191***	0.0000533***	0.000193***	0.0000538***
	(0.0000161)	(0.0000107)	(0.0000160)	(0.0000108)
2*median XS	–0.107***	–0.120***		
	(0.0179)	(0.0117)		
LIHC			–0.122***	–0.167***
			(0.0219)	(0.0159)
Constant	7.904***	8.224***	7.899***	8.218***
	(0.184)	(0.115)	(0.185)	(0.115)
Observations	100534	313838	100534	313838
R-squared	0.082	0.090	0.082	0.090

Note

Standard errors in parentheses. Significance: * 10% ** 5% *** 1%. Dependent variable: Life satisfaction (0–10 from not satisfied to very satisfied). All regressions control for individual socio-economic characteristics, individual, region, and time-fixed effects.

Table 10.3b Differences in mentality regarding energy poverty and income poverty

	E (East)	*F (West)*	*G (East)*	*H (West)*
Poor health	–0.467***	–0.497***	–0.468***	–0.497***
	(0.00918)	(0.00519)	(0.00919)	(0.00519)
Unemployed	–0.494***	–0.478***	–0.494***	–0.478***
	(0.0241)	(0.0202)	(0.0240)	(0.0203)
Equivalized income	0.000151***	0.0000445***	0.000154***	0.0000454***
	(0.0000159)	(0.00000931)	(0.0000158)	(0.00000944)
2*median XS	–0.0791***	–0.0878***		
	(0.0181)	(0.0115)		
LIHC			–0.0806***	–0.115***
			(0.0224)	(0.0159)
Income poverty	–0.201***	–0.198***	–0.201***	–0.192***
	(0.0225)	(0.0156)	(0.0227)	(0.0158)
Constant	8.045***	8.285***	8.040***	8.278***
	(0.186)	(0.115)	(0.186)	(0.115)
Observations	100534	313838	100534	313838
R-squared	0.084	0.091	0.084	0.091

Note
Standard errors in parentheses. Significance: * 10% ** 5% *** 1%. Dependent variable: Life satisfaction (0–10 from not satisfied to very satisfied). All regressions control for individual socio-economic characteristics, individual, region, and time-fixed effects.

chance to compare the effect of income poverty and energy poverty regarding the differences between East and West Germany. Columns A and C from Table 10.3a and columns E and G from Table 10.3b show the well-being effect of the 2*median XS measure, while columns B and D from Table 10.3a and columns F and H from Table 10.3b show the well-being effect of the LIHC measure.

Defining energy poverty as an individual form of deprivation that affects people's everyday lives, we expected the result that energy poverty is negatively correlated with individual life satisfaction. The negative effect is significant at the 1 percent level in all our models. This is in line with the findings from Biermann (2016). We also find the effect being different from mere material/income deprivation as the results are robust in Table 10.3b.[7] In addition to Biermann (2016) we see that energy poverty is significantly related to life satisfaction in East and West Germany. This is also what we expected. All other control variables are in line with those from life satisfaction studies using SOEP data as well (Dolan et al., 2008).

The more interesting result shows up comparing the results between East and West Germany. Let us recall the shares of unemployment, income poverty, and energy poverty from Table 10.1 and Table 10.2. The shares of unemployment, income poverty, and energy poverty are higher in East

Germany. The negative effect of unemployment and income poverty is greater (absolutely) in East Germany than in West Germany. This is in line with the argument that both unemployment and income poverty are classified as bad, where a higher share means a higher risk of being in the situation and, therefore, a greater negative effect on life satisfaction even for people not in the situation (Frey and Stutzer, 2002; Welsch and Biermann, 2019). This does not hold for energy poverty in this case. The respective results can be revealed comparing columns A, C, E and G with columns B, D, F and H, respectively. While the rates are higher (Table 10.1), the effect of energy poverty on life satisfaction is lower in East Germany than in West Germany. To explain the contrary results in terms of the rate–effect relationship, we need to discuss the causes of the negative effect of life circumstances, such as unemployment, income poverty, and energy poverty, on life satisfaction. For our explanation (in the German context) we are focusing on two mechanisms of how situations of deprivation affect life satisfaction. The existence and feasibility of coping strategies is one such mechanism. If people can cope more easily with their deprived situation, the effect on life satisfaction is lower. The other mechanism is related to stigmatization of marginalized groups. One of the mechanisms of how poverty affects life satisfaction is how the affected households are stigmatized. While coping strategies can be used to explain the difference between the results on unemployment and energy poverty, stigmatization is used to explain the different results on income poverty and energy poverty.

Households in income poverty are stigmatized by social prejudice, for example for being less educated, for having many children, and by a kind of socially deprived heritage from their parents' generation (Waxman, 1977; Reutter et al., 2009). In addition, people from this social "class" feel themselves to be stigmatized by others (Mickelson and Williams, 2008; Williams, 2009). By means of stigmatization, energy poverty should be worse for people in East Germany, because the energy poverty rate is higher. But, the stigmatization because of energy poverty is almost nonexistent in Germany since energy poverty is not a recognized problem in itself. Although it explains why energy poor households are better off in East Germany, this argument describes a trade-off in the field of energy poverty research and policy. If energy poverty becomes part of the official political agenda, it will be more recognized and targeted in redistributive policies, while this bears the risk of an increase (or an emergence) of stigmatization of energy poor households.

Another important mechanism is the existence of coping strategies (Anderson et al., 2012; Brunner et al., 2012; Chard and Walker, 2016). This is the part of the story which can best explain why energy poverty is even worse for households in West Germany, although the share of households affected by energy poverty is lower. Obviously, the main difference between East and West Germany is the socialist legacy from Soviet times in East Germany. The official and institutional structure is the same in both parts nowadays. There is no structural difference in terms of governance or administration. The coping strategies are performed at the individual level. In terms of heating expenditure, people in Germany have a similar coping strategy to people in other countries, which is mainly the use of fuelwood as a substitute. But in contrast to other countries this does not lead to issues like indoor air pollution, as there are relatively high regulative and technical standards for heating facilities (stoves). The acquisition of fuelwood has two types of suppliers and two types of channels. The suppliers are official market participants (supermarkets and retailers) on the one hand and semi-official suppliers on the other hand. The semi-official suppliers are private forest owners who – under some regulative restrictions – sell fuelwood based on private contracts. This is also the first channel of distribution for fuelwood in Germany (Wilnhammer et al. 2012; Jochem et al., 2015). There is a second channel which is based on informal distribution. These informal market structures are especially prominent in East Germany. Therefore, we expect coping strategies to be easier regarding the feasibility of informal transactions in East Germany.

We know that post-socialist societies have a higher prevalence of informal household exchange networks, sometimes called kinship-like systems, which we will refer to as kinship systems throughout (Pine, 1995; Acheson, 2007). They are not restricted to families and have a rather regional character in terms of kinship members. Within those kinship systems, households form informal distribution systems for goods and services. In East Germany, this unofficial distribution structure is further supported by private forest owners, representing the supply side of the kinship system. These systems are a legacy from the socialist past in East Germany. During Soviet times, they were necessary to be able to consume goods which were not produced in a manner that served high demand. Another factor supporting these coping strategies is the rural character of a large part of East Germany (relative to West Germany). East Germany has a relatively high share of rural population and a lower population density. The aforementioned kinship systems are more prominent in rural areas than in the more densely populated and economically more diversified and industry-shaped urban areas. Following these arguments, the existence of kinship systems in East Germany makes it easier to cope with energy poverty via the substitution of high energy prices through informal market structures. Since West Germany, in contrast, has a long history of a market economy, kinship systems are much smaller and not as pronounced as in the East.

Conclusions

We have shown that the differences in East and West Germany that are still prominent after 30 years of reunification also show up in the effect of energy poverty on people's well-being. In contrast to other deprivation categories such as income poverty or unemployment, people in East Germany are less affected by the negative consequences of energy poverty. We derived the thesis that this can (partly) be explained by a lower perception of energy poverty by the public and in politics as a separate issue, yielding a lower stigmatization effect for this marginalized group. Moreover, we hypothesized that as a result of the persistence of informal market structures (kinship systems) people in East Germany can cope more easily with energy poverty. The second part is especially interesting since it is a heritage from Soviet times and a part of the post-socialist character of East Germany. In many other countries in Eastern Europe, the post-socialist heritage negatively influences societies regarding energy poverty. One major difference in East Germany is the adoption of the West German constitutional, institutional, and regulative framework during the reunification process. Moreover, the infrastructure in East Germany has emerged as up to date (or even more modern) in relation to the Western part. Due to this unique situation described above, we have tried to find an explanation, not a blueprint, of an example where post-socialist heritage helps households on the individual level to cope with energy poverty, but only due to low stigmatization at the same time. This issue might become more prominent when energy poverty climbs the political agenda and becomes a recognized and publicly debated problem in Germany. Then it is questionable whether the relative position of energy poor households in the East will become worse than in West Germany, as is the case regarding unemployment and income poverty.

Notes

1 Differences in East and West Germany regarding life satisfaction, preferences, economic conditions, and cultural differences can be seen in, for example, Alesina and Fuchs-Schündeln (2007), Easterlin and Plagnol (2008), or Van Hoorn and Maseland (2010).
2 Due to data availability, Heindl (2015) concentrates on the expenditure-based measurement of energy poverty. There are also subjective measures and measures that account for psychological aspects. Here we focus on expenditure-based measures as well.
3 This method is explained in various econometric articles, see for example Galvao (2011).
4 Those individual socio-economic characteristics include marital status, age, education, health, and employment.
5 Regarding type 2 measures, we use two different measures since there is an ongoing discussion in the scientific literature on the 10% vs the 2*median XS measure, which we want to avoid in this context.
6 There is some debate about the "right" measure in the literature. We have opted for the 2*median XS measure, although some authorities and researchers are in favor of the 10% measure (Boardman, 2013; Schuessler, 2014). Concerning our empirical findings, we did a robustness check with the 10% measure, which confirmed our results for the type 2 measure (2*median XS).

7 In contrast to Biermann (2016) we split the sample between East and West Germany instead of running a regression on the overall population. When doing that as a robustness check, we found results confirming those in Biermann (2016).

References

Acheson, J. (2007). Household exchange networks in post-socialist Slovakia. *Human Organization*, 66 (4), 405–413.

Alesina, A., and Fuchs-Schündeln, N. (2007). Good-Bye Lenin (or not?): The effect of communism on people's preferences. *American Economic Review*, 97, 1507–1528.

Anderson, W., White, V., and Finney, A. (2012). Coping with low incomes and cold homes. *Energy Policy*, 49, 40–52.

Biermann, P. (2016). How fuel poverty affects subjective well-being: Panel evidence from Germany. *Oldenburg Discussion Papers in Economics*, 395 (16).

Biermann, P., and Welsch, H. (2019). Changing conditions, persistent mentality: An anatomy of East German unhappiness, 1990–2016. *Oldenburg Discussion Papers in Economics*, 422 (19).

Boardman, B. (2013). *Fixing Fuel Poverty: Challenges and Solutions*. Routledge.

Brunner, K. M., Spitzer, M., and Christanell, A. (2012). Experiencing fuel poverty: Coping strategies of low-income households in Vienna/Austria. *Energy Policy*, 49, 53–59.

Chard, R., and Walker, G. (2016). Living with fuel poverty in older age: Coping strategies and their problematic implications. *Energy Research and Social Science*, 18, 62–70.

Dolan, P., Peasgood, T., and White, M. (2008). Do we really know what makes us happy? A review of the economic literature on the factors associated with subjective well-being. *Journal of Economic Psychology*, 29 (1), 94–122.

Easterlin, R. A., and Plagnol, A. C. (2008). Life satisfaction and economic conditions in East and West Germany pre- and post-unification. *Journal of Economic Behavior and Organization*, 68, 433–444.

Ferrer-i-Carbonell, A., and Frijters, P. (2004). How important is methodology for the estimates of the determinants of happiness? *The Economic Journal*, 114 (497), 641–659.

Frey, B. S., and Stutzer, A. (2002), What can economists learn from happiness research? *Journal of Economic Literature*, 40, 402–435.

Galvao Jr, A. F. (2011). Quantile regression for dynamic panel data with fixed effects. *Journal of Econometrics*, 164 (1), 142–157.

Goebel, J., Grabka, M. M., Liebig, S., Kroh, M., Richter, D., Schröder, C., and Schupp, J. (2019). The German Socio-Economic Panel Study (SOEP). *Jahrbücher für Nationalökonomie und Statistik/Journal of Economics and Statistics* 239 (2), 345–360. doi: https://doi.org/10.1515/jbnst-2018-0022.

Heindl, P. (2015). Measuring fuel poverty: General considerations and application to German household data. *FinanzArchiv: Public Finance Analysis*, 71.2, 178–215.

Hills, J. (2012). *Getting the Measure of Fuel Poverty: Final Report of the Fuel Poverty Review*. http://eprints.lse.ac.uk/43153/.

Jochem, D., Weimar, H., Bösch, M., Mantau, U., and Dieter, M. (2015). Estimation of wood removals and fellings in Germany: A calculation approach based on the amount of used roundwood. *European Journal of Forest Research*, 134 (5), 869–888.

Mickelson, K. D., and Williams, S. L. (2008). Perceived stigma of poverty and depression: Examination of interpersonal and intrapersonal mediators. *Journal of Social and Clinical Psychology*, 27 (9), 903–930.

Pine, F. (1995). Kinship, work and the state in post-socialist rural Poland. *Cambridge Anthropology*, 47–58.

Reutter, L. I., Stewart, M. J., Veenstra, G., Love, R., Raphael, D., and Makwarimba, E. (2009). "Who do they think we are, anyway?": Perceptions of and responses to poverty stigma. *Qualitative Health Research*, 19 (3), 297–311.

Schuessler, R. (2014). Energy poverty indicators: conceptual issues-part i: the ten-percent-rule and double median/mean indicators. ZEW-Centre for European Economic Research Discussion Paper (14–037).

Van Hoorn, A., and Maseland, R. (2010). Cultural differences between East and West Germany after 1991: Communist values versus economic performance? *Journal of Economic Behavior and Organization*, 76, 791–804.

Wagner, Gert G., Joachim R. Frick, and Jürgen Schupp. (2007). *The German Socio-Economic Panel Study (SOEP): Evolution, Scope and Enhancements*. https://papers.ssrn.com/sol3/papers.cfm?abstract_id=1028709.

Welsch, H., and Biermann, P. (2019). Poverty is a public bad: Panel evidence from subjective well-being data. *Review of Income and Wealth*, 65 (1), 187–200.

Williams, W. R. (2009). Struggling with poverty: Implications for theory and policy of increasing research on social class-based stigma. *Analyses of Social Issues and Public Policy*, 9 (1), 37–56.

Wilnhammer, M., Rothe, A., Weis, W., and Wittkopf, S. (2012). Estimating forest biomass supply from private forest owners: A case study from Southern Germany. *Biomass and Bioenergy*, 47, 177–187.

Waxman, C. I. (1977). *The Stigma of Poverty; A Critique of Poverty Theories and Policies*. Pergamon Press, Inc., New York.

11 Regional disparities as roots of energy poverty in Slovakia

Dušana Dokupilová and Richard Filčák

Introduction

Most of the literature and scientific knowledge on energy poverty are derived from analyses at the level of nation states; this is even more the case for the study of the phenomenon in post-communist states, where few analyses exist at a regional level (for example Bouzarovski and Tirado Herrero, 2017). The present chapter shows that approaches that look at energy poverty only at a national level may hide regional disparities in the degree of energy poverty. This is visible in Slovakia, where, for instance, people working in the region around the capital city of Bratislava (southwest) generally earn average salaries that reach 160 percent of the average national wage. In contrast, in regions in the south and east of the country, salaries hardly reach 75 percent of the national average. Such striking differences are not uncommon across regions in post-communist states (Bouzarovski, 2018); thus, Slovakia is a good example to illustrate the link between regional disparities and energy poverty. Moreover, Slovakia illustrates a paradox of energy poverty: poorer people sometimes have to pay more for energy than the wealthier do (Filčák, 2010; Filčák and Živčič, 2017).

This chapter is divided into three main parts. First, we provide an overview of how the past three decades have led to regional disparities along dimensions linked to energy poverty. In the second section, we examine some of these dimensions to illustrate how they tend to concentrate along similar regional patterns in a manner that not only generates very high disparities and inequality, but also influences the concentration of energy poverty. The third section looks at recent policy developments and discusses their effectiveness.

The post-communist transition in Slovakia and the changing framework of energy poverty

The transformation of former centrally planned economies into liberal market societies led to deep and fast changes. Almost overnight, economies had to re-orient toward a new form of ownership, while opening to global competition. This transition occurred in a context of limited general knowledge about the pros and cons of the market economy, weak social resistance or, at least, minimal scrutiny, and public opinion demanding quick and effective crosscutting changes. This created what Leszek Balcerowitz considered to be a window of opportunity for "extraordinary policies" (Balcerowitz, 1995).

In the beginning of transition, the Slovak economy was characterized by a rather weak performance and obsolete economic structures compared to its Western counterparts. The dominant sectors were heavy industry and textiles, pointing at an economy similar to the earlier stages of industrialization. Knowledge-based sectors and information technologies lagged behind those of Western counterparts. Firms lacked flexibility and the marketing and other skills necessary to survive in a highly competitive environment. They also lacked experts in financial planning. Productivity was low, as often was the quality of products. Manufacturing was highly energy intensive and resource inefficient. In this context, the first wave of transformation meant a collapse of entire industrial branches, a rapid de-industrialization, and a decline in the production of agricultural goods. Regions suffered the impact in various ways, some taking the hit harder than others, a dynamic with effects that lasted a decade.

An economic revival became possible with a gradual increase in foreign direct investments and the 2004 EU accession. But the positive effects again displayed tendencies to concentrate unevenly geographically, being visible mainly in the western part of the country, partly due to the proximity to Western EU markets, and partly due to better infrastructure. A particular case is the Bratislava region, which ranks above the EU average in terms of economic development and living standards. This region has attracted most of the foreign investments and is a hub for companies with a national scope. At the level of the country, the internal migration of people follows the distribution of incomes and, in Slovakia, this migration is predominantly from the eastern and southern parts to the western part of the country; this dynamic further deepens regional disparities. Although Slovakia has recently reached a stable rate of economic growth, it is still failing to address these regional disparities.

Simultaneous to economic changes, the Slovak Republic went through gradual reforms of its welfare policies. The social protection system has been affected by a shift from welfare to the introduction of some workfare elements since the beginning of the 2000s.[1] The goal of building a more competitive, but also cheaper, labor force influenced the incomes and the purchasing power of households and put a lot of pressure on household budgets. Moreover, this dynamic has not been balanced by additional protection measures for vulnerable people.

These economic and social dynamics are deeply linked to the combination of drivers of energy poverty. During the post-communist transition, Slovakia underwent a massive privatization of the housing stock, resulting in a new generation of poor homeowners and tenants. At the same time, the prevailing housing stock – massive blocks of apartments built during centrally planned socialism – face wear and aging. Moreover, the property regime and the structure of the housing market, dominated by private property, requires a sharing of responsibilities between public and private owners, tenants and landlords. While in the wealthier regions, and especially in the capital city, the cost of houses is extremely high, in the poorer regions there is an increase in the number of abandoned apartments and prices are dropping. For instance, a two-room apartment built in the late 1980s may cost (as of January 2020) as much as 140,000 EUR in Bratislava, around 90,000 EUR in the central Slovak city of Banská Bystrica, and only 40,000 EUR in Humenné (eastern Slovakia).[2]

To summarize, robust economic and social reforms generated economic growth, but came with a decrease in social welfare measures. The disparities between the western and eastern regions illustrate the different development trajectories that can be observed within the country. Not even EU membership, which has undeniably brought relatively solid economic growth, has changed this trend of increasing social and regional inequalities. In the next section, we show that four dimensions associated with energy poverty are in close relation to these regional baselines.

An unequal spatial distribution of energy vulnerability factors

Slovakia is a country with 5.4 million inhabitants living in the territory of 49,036 km^2. It consists of eight administrative regions (NUTS 3 classification) and 79 districts (NUTS 4). Although Slovakia is a smaller European state, we notice major regional discrepancies on several dimensions associated with energy poverty (Gerbery and Filčák, 2014). The present section investigates four of these dimensions: income, geography and climate, age, and structural disadvantages and geographical spread of Roma communities.

Low incomes as a driver of energy poverty

Statistics related to economic productivity easily provide an image of regional disparities in Slovakia, where gross domestic product (GDP) per capita varies significantly across the country. Most of the production is concentrated in big cities, especially in the capital city of Bratislava. While the national average GDP per capita is 14,956 euro per year, in the richest region – that of the capital city – it reaches 36,704 euro. In the poorest region in the northeast of the country – Prešov – it is only 9,338 euro. The GDP is closely linked to wages and the employment rate. As seen in Figure 11.1, the poorest regions are located in the south and east of the country.

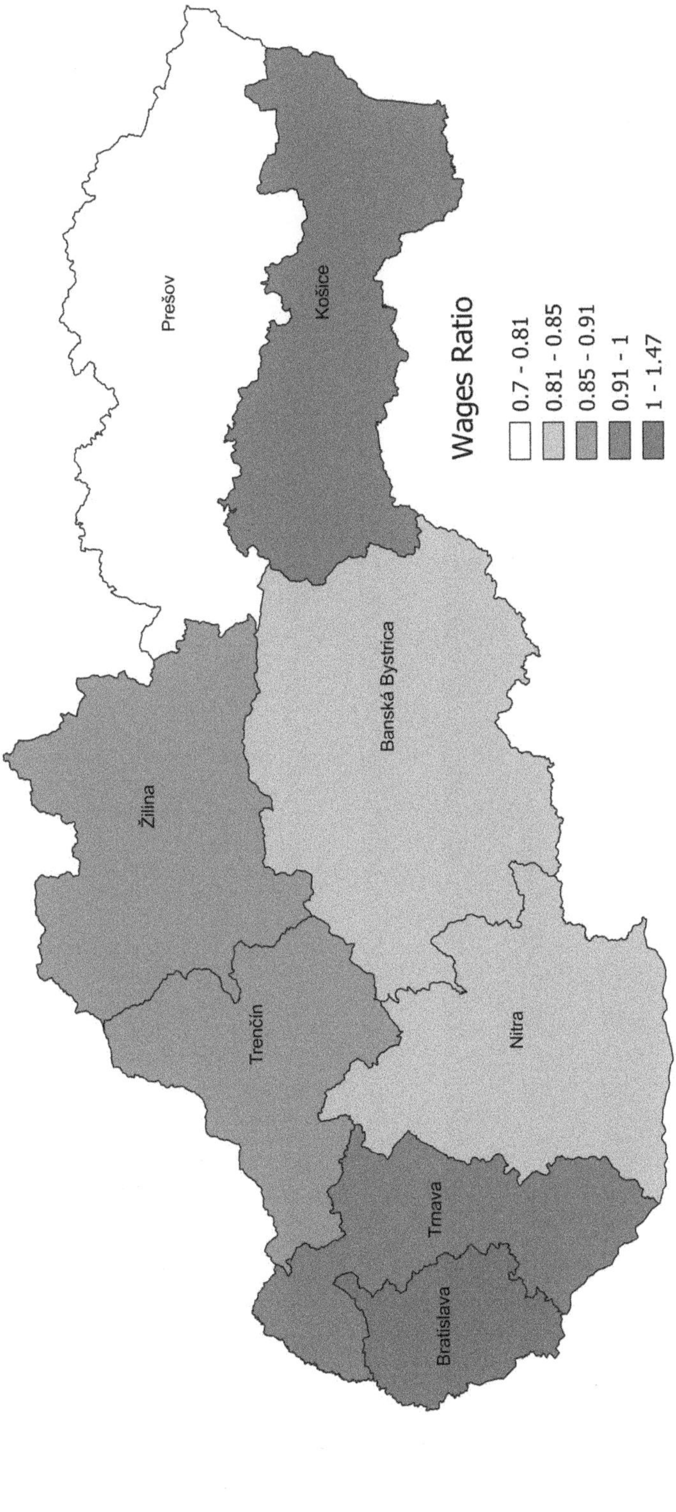
Prešov
Košice
Banská Bystrica
Žilina
Trenčín
Nitra
Trnava
Bratislava
Wages Ratio
0.7 - 0.81
0.81 - 0.85
0.85 - 0.91
0.91 - 1
1 - 1.47

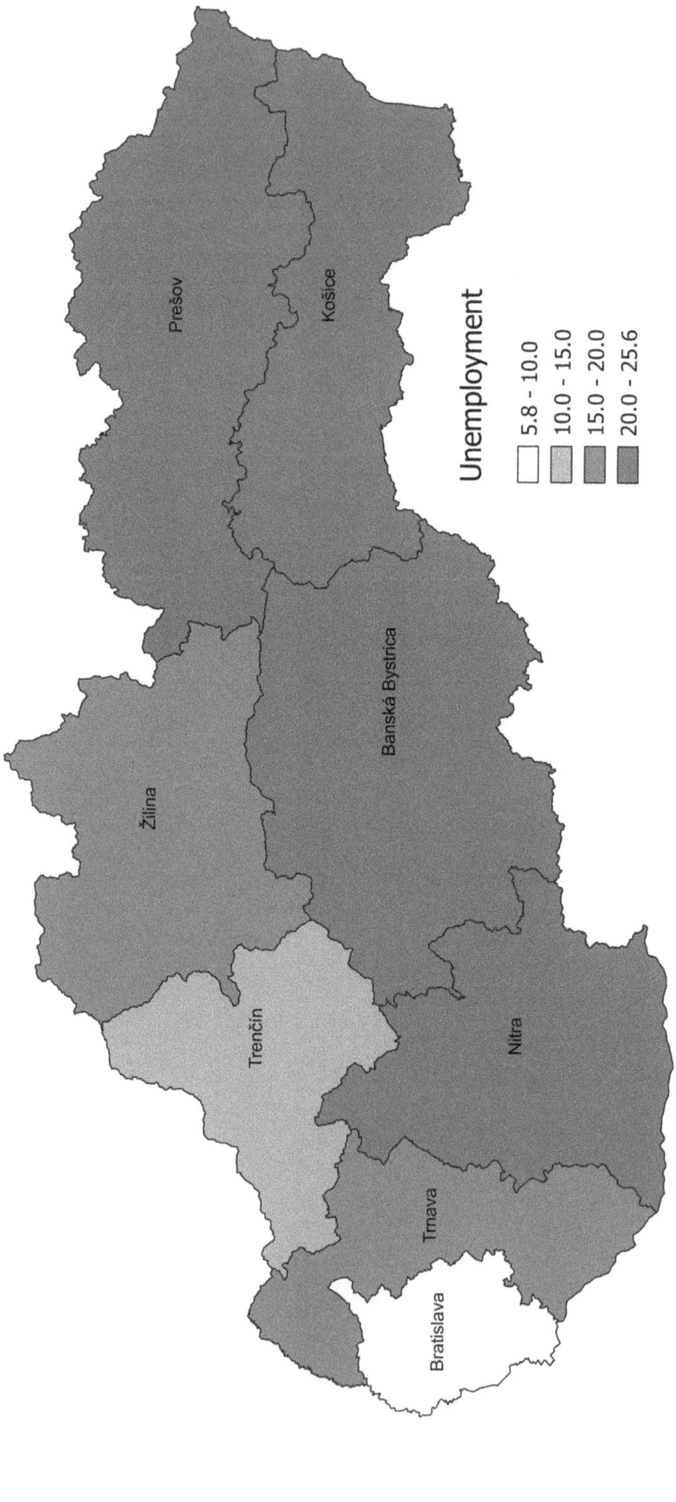

Figure 11.1 The average wage in the region compared to the average wage in Slovakia (ratio) (a) and the level of unemployment in regions (percent) (b) 2017.

Source: Authors' computation on data from Statistical Office of Slovak Republic.

The highest wages are in the biggest cities, such as Bratislava and Košice (southeast), where the average wage reaches 150 percent of the national average wage (Figure 11.1). On the contrary, the lowest average wages are in the peripheral regions of southern, central, and northern Slovakia, which reach less than 80 percent of the average national wage. The relatively good performance of the Košice region is determined by Košice city, the second biggest city in the country, while in the remainder of this region the situation is similar to other lower-wage areas.

Significant regional disparities are observed in the unemployment rate too (Figure 11.1). The lowest unemployment rates, below 3 percent, are in the districts around the capital city, in the west and in central Slovakia in the vicinity of large cities such as Banská Bystrica, and in the east in Košice. The highest unemployment rate is reported in the eastern and the southern regions of central Slovakia, where the unemployment rate, for instance in the district of Rimavská Sobota, reaches 2.59 times the average rate of Slovakia. Here, every fourth person is unemployed.

Despite the officially declared cohesion policy goals aiming to mitigate regional disparities, the situation did not change much over time. Until 2008, there was a gradual decline in unemployment. The decrease was sharper in regions with already high unemployment, which was also reflected in a decline in regional disparities. After 2008, however, districts with high unemployment saw sharper increases in unemployment while districts with lower unemployment were less affected, resulting in a rewidening of regional unemployment disparities. Since 2013, there has been a slight reduction of these trends.

The disparities in regional unemployment also translate into regional wage disparities, with the areas suffering from the highest unemployment also displaying the lowest wages. The combination of high unemployment and low incomes for those who are unemployed amplifies regional disparities. To illustrate the degree of inequality in terms of income at its peak, in 2009, the minimum wage for the city of Bardejov was 219 euro, while the highest minimum wage, in Bratislava II, was 638 euro. A decline in regional inequalities occurred after 2009. By 2012, the difference between the regions fell to 284 euro, with the highest incomes still in Bratislava II and the lowest still in Bardejov. Gradually, however, these inequalities are on the rise again.

Moreover, as Figure 11.2 shows, the number of people in material need also varies greatly by region.[3] The richest region, around Bratislava, has 11 times fewer people in material need than the poorest region. Correlation between the economic status of a region and the number of people in material need is clearly present in the context of the Slovak republic.

Climate and energy needs

Another important factor in understanding the regional dimension of energy poverty in Slovakia is geography, which is associated with climate.

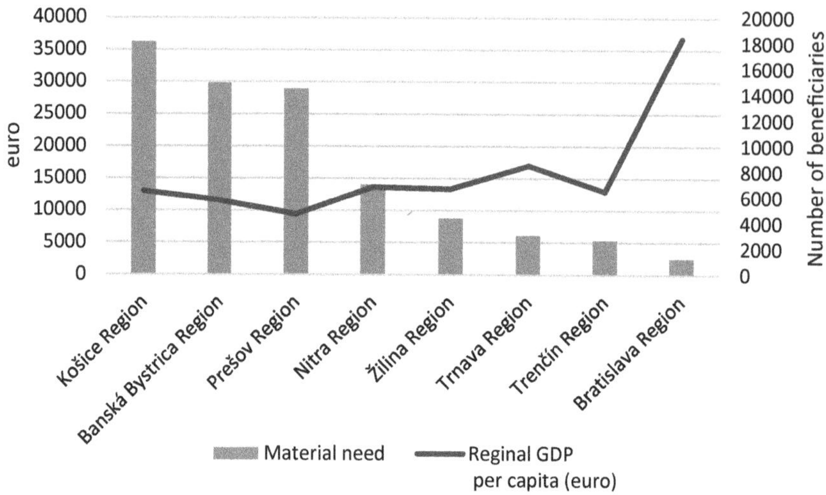

Figure 11.2 Regional gross domestic product per capita (in current prices) in comparison with the number of people in material need.

Source: Authors' computation on data from Statistical Office of Slovak Republic.

Slovakia is located in the middle of Europe and covers an area of less than 0.5 percent of the European Union. Its climate is influenced partly by the ocean, in the west, and is partly continental, in the east. Despite the small area, the altitude varies from 94 to 2,654 meters. This topography is closely linked to air temperature range – for example, in July (the hottest month in Slovakia) the average temperature varies between 4°C (in the mountains in the north of the country) and 20°C in the southern lowlands. On the other hand, the number of days when temperature during the day drops lower than 0°C is only 30 in the south and peaks to 190 days in the northern mountains. These conditions are major determinants of the quantity of energy needed to keep a household warm. Due to lower temperatures and longer winters, much higher consumption of energy during cold months can be expected in the northern parts of the country.

According to Slovak technical norms, district heating should be provided if the outdoor average daily air temperature drops below 13°C during two consecutive days and, according to the weather forecast, an increase of the outdoor average daily air temperature cannot be expected. Official statistics show that the highest number of district heating days are recorded in Brezno (a mountain city in the northern part of central Slovakia) and Tvrdošín (a city in the north of the country) – both with about 255 days. Districts like Kežmarok (north) and Gelnica (east) reach 243 heating days. Other places scoring high in terms of heating demand are Revúca and Sabinov with 225 heating days. The smallest number of heating days – about

190 – was recorded in Pezinok (Bratislava) and Senec, two cities located in the southwest of the country.

This adds up to the fact that the districts with the highest number of heating days are among the poorest in the country, while those with the smallest number of required days for heating are also economically rich (as in the area around the capital of Bratislava). This results in a high burden on energy budgets of households located in the coldest regions, with significant variations depending on household structure.

Old age and energy vulnerability

Age is a strong predictor of energy poverty in Slovakia. Data show how incomes and distribution of expenditures vary with age, with significant differences between people of productive age (employees) and retired pensioners (Table 11.1). An average household spends the highest portion of its expenditure on food. For households where the "head" is employed, 15 percent of expenditures go toward food. When the head of the household is retired, food expenses represent as much as 30 percent of expenditures. The second most important category of expenditures of all households is energy bills. An average household with an employed head pays 6 percent of its total expenditures toward energy. A household of retirees spends 15 percent of its total expenditures on energy services.

The total average expenditures of Slovak households where the head is employed are 597 euro per month. The average income of an employee in nominal values per month was 1,011 euro in 2018, decreasing to about 800 euro after deducting taxes and social expenses. This indicates that an "average" household can safely cover its expenditures with its monthly income. But many have incomes lower than what is required to cover all their basic needs. Approximately 4 percent of the population earns only the minimal wage, which is readjusted every year. For 2019, it was set to 520 euro per month, decreasing to 430 euro after deducting taxes and social expenses. Thus, a person receiving the minimal wage has insufficient financial resources to cover the average expenditures.

Retired people make up a significant group that is vulnerable to energy poverty. In Slovakia, over 20 percent of the population is retired. The average pension is 459 euro per month. The distribution of pension monetary values is almost statistically normal, with only a small number of people reaching extreme values (Figure 11.3). Obviously, the group with a very low pension, below 200 euro per month – 1.7 percent of retirees (18,460 persons) – is very far from reaching the average needs of a retiree household, which, according to Table 11.1, is quantified at 356 euro. Based on data from the social insurance company, which is the state institution that pays pensions, almost 194,000 people (22 percent of retirees) have a pension below 356 euro per month. These people need to use their savings or find additional sources of income by working, renting a part of their dwelling, or asking relatives for financial aid.

Table 11.1 Household expenditures according to head's retirement status in 2018 (euro/month/person)

	Employee's household	*Share of the item in total costs (%)*	*Retiree household*	*Share of the item in total costs (%)*
Food and non-alcoholic beverages	75.96	13	95.39	27
Alcoholic beverages and tobacco	9.56	2	10.23	3
Clothing and footwear	19.42	3	10.48	3
Housing	25.83	4	27.85	8
Energy	36.11	6	52.9	15
Furniture, furnishings, and routine maintenance of the apartment	18.08	3	15.79	4
Health	9.45	2	17.45	5
Transportation	45.37	8	23.28	7
Post and telecommunications	17.97	3	14.86	4
Recreation and culture	23.77	4	16.57	5
Education	2.84	0	0.35	0
Hotels, cafes, and restaurants	20.81	3	11.06	3
Other gross expenditure	182.94	31	48.92	14
Income tax	43.94	7	2.21	1
Payments to health insurance companies and social insurance companies	61.12	10	4.48	1
Consumption of food from own production and donations	3.89	1	4.42	1
Total expenditures	597.06	100	356.24	100

Source: Statistical Office of the Slovak Republic.

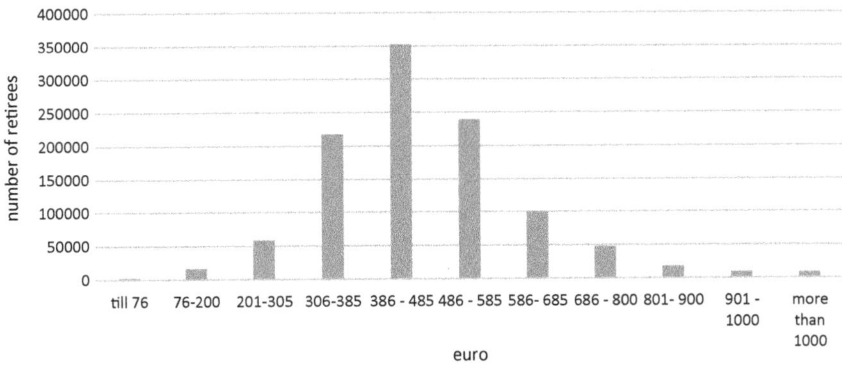

Figure 11.3 Number of retired people per pension amount in euro 2018.

Source: Authors' computations based on data from the Social Insurance Company of Slovakia.

The public debate on energy poverty revolves around setting a threshold of at least 10 percent of expenditures going toward energy needs, as we detail later in the chapter. What does this mean for retirees? With the average expenses for energy at 52.9 euro, incomes below 529 euro would qualify as leaving a retiree household in energy poverty. In Slovakia, over 72 percent of retirees (770,000 people) have pensions below this threshold. However, the proposed definition ties this 10 percent threshold to an additional requirement of being identified as being in material need and that has a different, even more restrictive, threshold. This combination would reduce the number of retirees which might get some kind of aid for energy expenses to only 18,400 persons, meaning only 1.7 percent of retirees and 0.4 percent of total population.

Vulnerable communities: the Roma

A specific group extremely exposed to energy poverty in Slovakia is the Roma population. The Roma represent an estimated 8 percent of the country's total population, with the biggest concentrations of Roma settlements found in the poorer regions of the east and south of the country (UNDP, 2014; Atlas, 2019). The vast majority of Roma now live integrated in towns and villages, yet part of the minority still lives segregated away from concentrations of dwellings, often without access to basic services, in extreme cases even without electricity or drinking water. The early years of economic post-communist changes had serious consequences for this minority. Many Roma lost their jobs, while the privatization of the housing stock led to many cases of eviction leading to Roma often being forced to move from their original dwellings to segregated settlements at the outskirts of villages or towns.

Statistical surveys are sometimes criticized for insufficient coverage of the most vulnerable groups that live at the margins of society. Marginalized Roma households represent one such example. In order to capture their living conditions and provide a comprehensive overview for policy-makers, the United Nations Development Programme (UNDP), with the support of nongovernmental organizations (NGOs) and academics, conducted the Survey on Living Conditions of Roma Households in Slovakia, in 2010 (UNDP, 2014). The survey provided data on a broad range of issues. Access to energy and energy poverty are part of the complex context that shapes the lives of the Roma. The UNDP study reports in particular a high incidence of overcrowding (26 percent of Roma households live in dwellings with a maximum of 5 square meters per person), of inadequate access to drinking water (6 percent of Roma households do not have access to drinking water), and non-standard conditions for personal hygiene. Almost every fifth Roma household surveyed reported an interruption of their electricity supply during the previous year. Out of this fraction, 38 percent were left without electricity for three or more months (UNDP, 2012, p. 71). The most frequent reason for disconnection from the electricity supply was a lack of financial resources (68 percent). Disconnection as a result of technical difficulties was rare (12 percent). This reveals a very deep pattern of exclusion from a decent standard of life for a significant part of the population. Evidence that most Roma households use non-standard devices for heating their dwellings supports these findings. Central heating is uncommon to Roma households (7 percent of households, compared to 48 percent of households belonging to the control group who live in proximity), for which wood/coal stoves represent the most common method of heating.

While in the more developed and warmer western part of the country, the Roma account for only 2–5 percent of the population, the highest proportion of the Roma lives in eastern Slovakia, where one in five citizens (over 18 percent) is Roma. At the same time, this area is one of the poorest, displaying the lowest GDP, the highest unemployment rate, and the lowest salaries. There is also a significant overlap between the location of Roma communities and the coldest regions within the country. Many dwellings have a very low level of energy efficiency, since they are made simply of wood, sheet tin metal, or other readily available materials. The cumulative outcome is that Roma, as a specific social group, are deeply exposed to living in energy poverty.

Resulting vulnerability of the regions to energy poverty

Based on the four dimensions associated with energy poverty on which we identify variation among the regions in Slovakia, we can conclude that, in Slovakia, some regions are more prone to energy poverty than others. In the Prešov region, for instance, the unemployment rate is high, income is low, some areas are among the coldest in the country, and it has many Roma communities. Moreover, the lower average age might in fact be due to the high proportion of Roma, who have lower average age than the rest of the population. On the other hand, the Bratislava region displays a higher average age of population, but on the other dimensions it ranks as the least vulnerable region in country. Even the climate is mild, since the southwestern part of the country with a medium to lower altitude does not require a lot of energy during winter. Furthermore, there are almost no segregated Roma communities in this region. A synthesis of this evaluation across all regions is displayed in Table 11.2.

Regarding the prices of fuels used for heating, there are some adjustments considering the location of the house relative to altitude, but the differences are not significant.[4] When looking at the levels of incomes and GDP per capita across regions, we conclude that households in the northern parts of Slovakia spend a higher portion of their income on energy and are thus much more exposed to energy poverty. This holds regardless of the structure of the household. These regional disparities are not accounted for in Slovakia's unified and centralized welfare system: therefore, policy responses to energy poverty are not adapted to the variations in regional contexts.

Another layer deepening the structural exposure of some regions to energy poverty is the overlap between the low quality of the housing stock and low incomes: not only do households in disadvantaged regions and with a higher demand for heating have to pay more for energy, but they also are more likely to lack the resources to invest in energy efficiency. Energy efficiency is a key factor that is both bound up with energy poverty and crucial to addressing it. In some regions of Slovakia, many people do not have enough money to cover their everyday life expenses – for example

Table 11.2 Vulnerability of regions to energy poverty factors

Region	*Vulnerability factors*			
	Low incomes	*Climate and energy needs*	*Old age*	*Vulnerable communities – Roma*
Bratislava region				
Trnava region				
Trenčín region				
Nitra region				
Žilina region				
Banská Bystrica r.				
Prešov region				
Košice region				

Source: authors' own elaboration.

in the Košice, Banská Bystrica, and Prešov regions. Moreover, these regions are located in the mountainous area with the longest and coldest winter, which results in high energy requirements.

In summary, the coldest regions like Prešov and Košice tend to show a concentration of people with low wages (leading to low pensions), Roma ethnic minority, and single parents, making them the areas facing significant energy-poverty-related problems. These regional patterns of population in conjunction with income distribution point to the importance of developing policies and projects to tackle energy poverty as a regional phenomenon.

Policy approaches and the regional dimension of energy poverty

Poverty, with its multidimensional aspects, has been for a long time at the center of public policies, but energy poverty, targeted as a specific issue, is a latecomer (Pye and Dobbins, 2015; Walker and Day, 2012). In Slovakia, the launch of official discussions about energy poverty and public policies addressing it dates back to 2012. These followed a public debate stimulated by several research studies and, most of all, by an increased awareness of the public regarding energy-related matters.

According to the Act on Regulation (250/2012) of 2012, the Regulatory Office for Network Industries (RONI) is obliged to draw up a "concept" for the protection of customers meeting the conditions of energy poverty. In 2016, this was set as a goal in the program of the (then) new government. The same year, RONI declared that Slovakia would adopt an energy poverty law by the end of the year. A working group was set up consisting of representatives of the Ministry of Economy, Ministry of Finance, and Ministry of Labour, Family and Social Affairs, as well as the Regulatory Office. The first legislative proposal to address energy poverty was prepared relatively quickly. Commenting on the draft concept ended on January 3, 2017. There were 101 amendments to the proposal, of which 49 were substantive. In 2019, RONI set up the draft of the Energy Poverty Concept for public discussion, which is still under discussion.

A very important part of this Energy Poverty Concept is the first attempt to define energy poverty in Slovakia in an official document. The proposed definition is:

> Energy poverty refers to a situation where individuals or households do not have enough funds to provide heating and other energy needed for the functioning of the household, which in the Slovak Republic represents 10.0 percent of the average household expenditure on energy from the total net money income of the household. And simultaneously the household meets the conditions for granting a benefit in material need.
>
> (URSO, 2019)

The concept was met with severe criticism, especially from the Ministry of Labour, Family and Social Affairs. The main concern is that the definition should lead to new social and financial measures, which would put additional pressure on the public budget.

In early 2020, attempts to attach energy poverty to the social welfare system ceased with the new general elections. It is currently unclear how the topic will be dealt with by the new government. In the electoral programs, only one political party included energy poverty in its program. This was a neoliberal civic party which only proposed introducing pre-paid cards for electricity for poor people.

Even in the absence of targeted and binding measures for energy poor households, some opportunities are already embedded in the current policy framework. According to Act 417/2013[5] of 2013, households are entitled to material assistance if the income of household members does not meet the minimum subsistence income stipulated by a special regulation, and household members are not working or are not able to work for an income, and do not own property or use other rights to property. In 2019, the minimum income was set at 210.20 euro per month per single adult person, plus 146.64 euro per month if there is another dependent adult person, and 95.96 euro per month per dependent child. The minimum thresholds are systematically criticized by academics and NGOs as being too low, which excludes a significant part of the population in need of social assistance. Inevitably, the low thresholds for qualifying incomes drive down the number of households that qualify as being in material need. Tying the thresholds for energy poverty to low incomes and material need would have similar consequences, by diminishing the number of households counted as energy poor and, consequently, targeted by subsequent policies.

Finally, it is important to note that the structure of the welfare system, which reflects the centralized administrative organization of the country, does not place emphasis on regional disparities, and therefore there are no tailored solutions depending on the problems at the local and regional level. The main measure aimed at identifying and targeting the poorest local communities is making a list of so-called least developed districts. There are currently 20 on the list, out of the 79 total in the country.[6] These districts are eligible for special measures and financial support, coming mainly from European cohesion funds. However, none of these funds are designed to tackle energy poverty, mainly since, despite the recent debates, the phenomenon is not yet officially legally acknowledged and defined.

Conclusions

Our analysis points to the importance of regional disparities along several dimensions as factors contributing to energy poverty. We believe any future strategy aimed at mitigating energy poverty needs to address them. The roots of the striking regional disparities may be traced back to rapid de-industrialization and the subsequent decline of certain regions in the early years of transition, coupled with the declining quality of housing and the massive privatization of the housing stock. This has created a generation of poor owners and tenants living in disadvantaged regions, with systematic difficulties in paying for their energy needs. More recently, despite the relatively good economic performance of the country, social and regional inequalities are growing in Slovakia, which might reinforce these regional disparities. Currently, the main challenge in addressing energy poverty centers around with its official definition. Having one that captures the issues at the heart of regional disparities would allow assistance and funds to be targeted more effectively.

Notes

1 Especially with the Act on Assistance in Material Need No. 599/2003 Coll. adopted in 2003.
2 See for instance Slovak National Bank analyses at: www.nbs.sk/sk/statisticke-udaje/vybrane-makroekonomicke-ukazovatele/ceny-nehnutelnosti-na-byvanie.
3 Material need is a situation where the income of household members is below the subsistence level. Additional information in Section 3.
4 The adjustment applies to natural gas only and is partly driven by the physical performance of the gas. Until 2008 gas consumption was measured per m^3. However, this measurement severely disadvantaged households at higher altitudes. Since atmospheric pressure is lower at high altitudes, it leads to a lower density of natural gas in the pipelines, followed by lower efficiency. Therefore, the conversion of gas consumption per kWh (based on the heat that the gas can produce) has been calculated. As a result, areas at lower altitudes have a conversion factor from m^3 to kWh greater than 1, while for municipalities at high altitude this coefficient decreases to 0.88.
5 On Assistance in Material Need and on Amendments to Certain Acts.
6 See the list at www.upsvr.gov.sk/statistiky/zoznam-najmenej-rozvinutych-okresov.html?page_id=561733.

References

Atlas rómskych komunít. (2019). Ministry of Interior of Slovak Republic. Available at: www.minv.sk/?atlas-romskych-komunit-2019.

Bouzarovski, S. (2018). "European energy poverty," policy brief. Available at: www.engager-energy.net/wp-content/uploads/2019/01/Engager-Brief-1.pdf.

Bouzarovski, S., and Tirado Herrero, S. (2017). "Geographies of injustice: the socio-spatial determinants of energy poverty in Poland, the Czech Republic and Hungary." *Post-Communist Economies, 29*(1), 27–50.

Filčák, R. (2010). "Concept of 'energy poverty' and multi-dimensional perspectives of social inequalities and their impacts: case of the Czech and Slovak Republic." In: *Energy for Sustainable Development*, Haas, R., Jílkova. J. (editors). Praha: Alfa Press, ISBN: 9788087197363.

Filčák, R., and Živčič, L. (2017). "Energy poverty and multi-dimensional perspectives of social inequalities and policy challenges." *International Issues and Slovak Foreign Policy Affairs, 26*(1–2), 40–61.

Gerbery, D., and Filčák, R. (2014). "Exploring multi-dimensional nature of poverty in Slovakia: access to energy and concept of "energy poverty." *Ekonomický časopis*, 6, 579–598.

Pye, S., and Dobbins, A. (2015). "Energy poverty and vulnerable consumers in the energy sector across the EU: analysis of policies and measures." Published by Insight-E. Available at: https://ec.europa.eu/energy/sites/ener/files/documents/INSIGHT_E_Energy%20Poverty%20-%20Main%20Report_FINAL.pdf.

SOSR. Statistical Office of Slovak Republic. Available at: http://datacube.statistics.sk/.

UNDP. (2014). Atlas rómskych komunít na Slovensku 2013. Available at: www.minv.sk/?atlas_2013.

ÚRSO. (2019). návrh Koncepcia na ochranu odberateľov spĺňajúcich podmienky energetickej chudoby. Available at: www.slov-lex.sk/legislativne-procesy/-/SK/LP/2019/332.

Walker, G., and Day, R. (2012). "Fuel poverty as injustice: Integrating distribution, recognition and procedure in the struggle for affordable warmth." *Energy Policy*, 49, 69–75.

12 Conclusions

Energy poverty as a threat to democracy in post-communist countries

George Jiglau

In their seminal book on the path and challenges of transition to democracy in Central and Eastern Europe (CEE), Linz and Stepan (1996, pp. 3–6) make important conceptual distinctions between "democratization," "liberalization," and "democratic consolidation," which are key to understanding the profound transformation occurring in post-communist states since 1990. Even when the traits of democratic consolidation are present, they argue:

> We do not preclude the possibility that at some future time it could break down. But we do mean to assert that such a breakdown would not be related to weaknesses or problems specific to the historic process of democratic consolidation per se, but to a *new* dynamic in which the democratic regime cannot solve a set of problems, a nondemocratic alternative gains significant supporters.
>
> (Linz and Stepan, 1996)

Post-communist states in CEE are not the only democracies faced with a rise in populism in the past decade. From a practical perspective, many have already acquired the consolidated label as a result of NATO enlargement and EU accession, while others are gradually entering the negotiation process or at least preparing to begin it. But post-communist democracies are still young compared to their counterparts across the democratic Western world. In these counterparts, too, many democratic regimes face the rising challenge of populism, with all the many vicious circles revolving around a mix of social, economic, political, and institutional issues associated with it. However, the risk of democracy backsliding into illiberalism – a phrase that became rooted in the literature (McCormick et al., 2019) and is regarded as an intermediate step before the "democratic breakdown" theorized by Linz and Stepan (1978; 1996) – has become obvious. This is due either to a lack of sound checks and balances inserted in post-communist constitutions, or to the ease with which certain leaders or parties topple constitutions and increase their control in the absence of powerful political or civil society opposition (Bieber et al., 2018; Mudde 2013).

There are few phenomena which are as crosscutting as energy poverty. The substantial and still-developing literature thrives on multi-disciplinarity and has come a long way from the early approaches that linked the inability to find adequate thermal comfort in a household only to material poverty and the incapacity to buy energy or pay for it. We now know that energy poverty is the cumulative result of shortcomings along multiple dimensions of economic and social life, market structures and property regimes, policy-making and actions of private agencies, and (changing) climate and relief structure (Sareen et al., 2020, Dubois' Introduction in this book). Beyond the literature, we see the multi-dimensionality of the phenomenon in scientific research endeavors as the Energy Poverty Observatory, which collects data on a wide array of indicators, or international scientific networks, such as the ENGAGER COST Action. This book makes no exception; among the 22 authors and editors contributing to this book, we have economists, anthropologists, engineers, mathematicians, political scientists, sociologists, and philosophers.

This diversity makes energy poverty an almost permanently compelling phenomenon to study, with the particular feature of having a direct connection to policy-making. This is why many in the field thrive on the agenda as there is substantial evidence that their academic effort ultimately has an impact on policy-makers – albeit there is still room for improvement in this chain of communication. Many of the policies passed at local, national, or EU level in recent years are grounded in scientific work. We can safely argue that research on energy poverty ultimately has a positive impact on people's lives.

However – moving back to the features of post-communist states – it is the same multi-dimensionality that explains why energy poverty is so deeply rooted in the region. Essentially, the purpose of this book has been to tell the story of how energy poverty became noticed in relation to post-communist legacies at the macro and micro levels across CEE, but also to provide a comprehensive perspective on the policy responses given to the phenomenon. The chapters display both an array of similarities and variation in the attempts to acknowledge and address energy poverty. But one major thing that all countries have in common, visible even in the absence of unequivocal statistics, is that the policy responses of authorities are absent, inadequate, mistargeted or, at best, incomplete. Energy poverty may be slowly being acknowledged and addressed, but it is also perpetuating. Could this be the "*new* dynamic in which the democratic regime cannot solve a set of problems" which might lead to – or, at least, help – "a nondemocratic alternative gain significant supporters" (Linz and Stepan 1996, p. 6)?

In phrasing this, I am aware that nondemocratic alternatives already have plenty of supporters.

Populist politicians do not openly take on "energy poverty" as a phenomenon: they do not mention it in their discourses, and do not build political campaigns on it. The topic is barely perceived by voters themselves, so it is difficult to make a political *thing* out of it. However, we can discover dimensions of energy poverty already attached to successful populist discourses. Once you have read the chapters of this book, I encourage you to go back and re-read them with this fresh perspective in mind. Hungary's story highlights the politicization of energy prices by Viktor Orban's Fidesz in the 2014 and 2018 electoral campaigns, while the government's reshaping of the welfare system emphasizes and rewards the merits of employed and well-behaved citizens, further trapping poorer citizens in a vicious circle (see p. 000). In Romania, a highly controversial emergency (implicitly non-transparent) decree by the government at the end of 2018 was passed with the official intention – among many other interventions in the economy – to cap energy prices, but in fact did so only for electricity at the cost of generating discrete destabilization in the overall structure of the energy markets. Nevertheless, the government and the governing party claimed that these are measures to protect vulnerable consumers, basically misleading the public (see p. 000). The chapter on Slovakia, with its accent on regional disparities, points to the deep link between energy poverty and the situation of the Roma minority, particularly burdened by a mix of poor housing, unemployment, and low incomes (see p. 000). One of the regions where this link is most visible is Banská Bystrica, where far-right politician Marian Kotleba was elected governor in 2013 after an aggressive campaign that stigmatized the Roma. Furthermore, in 2016, Kotleba and his party won above 8 percent in the parliamentary elections and he later won above 10 percent in the presidential elections of 2019. He continues to poll well especially in Banská Bystrica and the poorer areas in central Slovakia. Part of his agenda exploits the complexity of pricing mechanisms for energy, which he quickly orients into the promise of nationalizing the energy system (Radek, 2019).

The structure of energy markets and the ongoing process of liberalization, especially as a result of EU membership, raise many opportunities to fuel populism, as it feeds on fingers pointing at too much complexity that allegedly hides the corrupt and self-interested practices of the main players, and is detrimental to ordinary citizens (Mudde, 2004). Energy markets – even with the ongoing liberalization – are a mix of state regulation and action of private, profit-oriented entities. The state institutions (which include governments, parliaments, local authorities, and energy regulating agencies) have a solid degree of politicization. Both these state institutions and big private companies – "corporations" – are part of what is labeled an "elite," often in the negatively twisted sense of the word. This elite dominates, in terms of influence, energy markets everywhere, and generally suffers from low levels of trust from citizens. As opposed to other goods and services people buy, energy services are vital for any household; therefore, they are constantly under scrutiny and can easily generate discontent. This lays the groundwork for dissatisfaction with political institutions and private actors, which can easily be generalized and singled out as the "political and economic elite," an oversimplified enemy that is an easy target for populist discourses and, if electorally successful, populist policies. Energy market structures and the high levels of dissatisfaction and mistrust are common across the entire post-communist space (and even beyond). To add to the complexity, fuel supply – especially with respect to natural gas – is heavily dependent on international suppliers, adding yet another layer of complexity to political and private actions which ultimately influence the price of fuel, as the chapters on Romania and Lithuania point out.

Beyond energy markets, the structure of the housing market and the state of the housing stock, in a strong connection to property regimes, are also a common elements across the post-communist space (perhaps with the exception of Czechia and Germany) and present many challenges associated with energy poverty (Bouzarovski and Tirado Herrero, 2017). As the chapters on Romania, Lithuania, Hungary, and Slovenia point out, the massive privatization of the housing stock in the early stages of transition, coupled with characteristics from the communist way of planning, led to dwellings that are rapidly degrading and that require significant investments in order to increase their level of energy efficiency. Low income comes as a major hurdle to many private households, a reality which can be compensated for either by state intervention or by cheap lending opportunities mainly from banks, a dynamic illustrated by our chapters on Slovenia, Bulgaria, North Macedonia, and Poland. State intervention depends on local or central public budgets, which are rather limited or depend on long-term good economic growth and effective redistribution policies. On the other hand, banks are also private, profit-oriented entities; therefore, pricing their loan products below a level justifiable on business grounds requires either, again, state support, or a sense of community responsibility which is not always in a bank's culture.

In terms of policy-making, one common element across post-communist states is the linking of energy poverty, once it begins to be acknowledged, to low income. Essentially, in the policy-making realm, energy poverty is tied mainly to poverty in the first place. Only gradually is it being linked to air quality – as in Poland or Bulgaria – or to energy efficiency, and even less so to accessibility of various fuels, the quality of equipment inside the household, or behavioral aspects that have an impact on fuel consumption. Furthermore – mainly because of the centralized administrative models at work throughout the post-communist space – the attention paid to local specifics of energy poverty is limited, as our chapter on Slovakia points out. We run into a paradox: a highly complex phenomenon is being approached by policy-makers in simplistic ways. The initial solutions, most often through some kind of heating aid in the cold season, only function as blanket measures, but it quickly becomes obvious that they fail to address the root causes of the phenomenon (Bouzarovski and Tirado Herrero, 2017). Additional elements are then added, but the common outcome across all states is a set of policies that lacks coherence, as the chapter on Bulgaria emphasizes, leaving many gaps which allow for energy poverty to perpetuate. On the brighter side, another common element of policy-making mechanisms across all countries is the gradual inclusion of civil society actors and of socially minded researchers in the process – some included in this book, as it happens, for instance, with our contributors from Slovenia, Poland, and Romania. This constitutes an approach that eventually delivers better-integrated results, even if decisions are still ultimately made at an institutional level and there is still a long way to go until all shortcomings are eliminated. However, this also adds up to the complexity populists can pick on. Civil society – mainly nongovernmental organizations – has taken on a bad connotation in populist discourses (for a perspective on the case of Hungary, see Antal 2019). Its involvement in the mix of stakeholders dealing with energy poverty, along with politicians and private actors, rounds up the populist image of an elite that does not properly address the problems of citizens.

The reflex of making energy poverty a social issue is facilitated by the fact that, everywhere, only certain categories of society are primarily affected. For instance, older people are vulnerable to energy poverty throughout the post-communist space, not only because of lower incomes, but also because of increasing expenses for medicine and medical services, lower access to information, difficulties in interacting with the relevant companies and authorities, and shrinking social networks. Roma communities, with very distinct features and present across all post-communist countries, are also deeply affected by energy poverty and face additional disadvantages stemming from the difficulties they have in interacting with authorities in order to access even the existing blanket measures. However, it is important to note that energy poverty is not limited only to some clearly defined social or demographic groups of people and the drivers of energy poverty, outlined in the introduction to this book (see p. 000) are often diffused within the population. For instance, all the chapters mention the need for significant upgrades to the housing stocks. This need for housing stock upgrades, coupled with the peculiar property regimes mentioned above, creates the premise for difficulties in preserving adequate thermal comfort at an affordable price without undue pressure on household incomes.

These are only some of the underlying issues raised by the chapters in this book. Throughout the post-communist space there is still a need to acknowledge, at a social and political level, that energy poverty is present, that it is the result of an accumulation of problems in multiple dimensions across time, and that tackling it efficiently requires long-term coherent policies and the participation of multiple stakeholders that must do their part in identifying those affected and in coming up with tailored solutions. The roots of energy poverty are deeply embedded in how communist regimes designed cities and villages, energy infrastructures, institutions, economies, and societies themselves. The early stages of transition failed to find answers and instead created additional challenges, like the de-industrialization of large areas or the slow adaptation to market economies and the associated high unemployment, inflation, and decline in purchasing power. All these made way for growing social inequality, and also a prerequisite of energy poverty.

The answers to energy poverty are not easy and it is not only post-communist countries that strive to find them. One of the purposes of writing this book has been to contribute to an additional layer of knowledge on energy poverty, not in terms of a geographical space, but rather on a political space, with very clearly defined boundaries and a specific set of features. CEE began the discussion on energy poverty – in the political, public, or academic space – roughly two decades later than Western Europe. With this in place, we hope there will be more room for comparative endeavors in the study of energy poverty, for better identifying how energy poverty unfolds, and, ultimately, for finding better politically sustainable solutions for those affected by the phenomenon – and, returning to the starting point of these conclusions, for finding these solutions as soon as possible. My point throughout this section of the book has been to emphasize that the rise in populism and illiberal tendencies that threatens democracies everywhere is already especially prevalent in Eastern Europe (Hungary, Poland, other countries to a lesser extent) compared to other democracies and that it has already begun to pick on energy-poverty-related issues. Energy poverty offers fertile ground for populist discourses, as it thrives on structural problems that are difficult to address and require the involvement of a multitude of stakeholders. Maybe the link between energy poverty and the stability of democratic regimes is not obvious and may seem far-fetched at first, but we need to be aware, before it is too late, that it exists and it is already being exploited.

References

Antal, Attila. 2019. *The Rise of Hungarian Populism: State Autocracy and the Orbán Regime*. Emerald Group Publishing.

Bieber, Florin, Magdalena Stolska, Dane Taleski. 2018. *Illiberal and Authoritarian Tendencies in Central, Southeastern and Eastern Europe* (Interdisciplinary Studies on Central and Eastern Europe). Peter Lang AG, Internationaler Verlag der Wissenschaften.

Bouzarovski, Stefan, Sergio Tirado Herrero. 2017. "Geographies of injustice: the socio-spatial determinants of energy poverty in Poland, the Czech Republic and Hungary," in *Post-Communist Economies*, Vol. 29, Issue 1, pp. 27–50.

Linz, Juan, Alfred Stepan. 1996. *Problems of Democratic Transition and Consolidation: Southern Europe, South America, and Post-Communist Europe*. The Johns Hopkins University Press.

Linz, Juan, Alfred Stepan. 1978. *The Breakdown of Democratic Regimes*. The Johns Hopkins University Press.

McCormick, John, Rod Hague, Martin Harrop. 2019. *Comparative Politics and Government* (11th edition). Palgrave Macmillan.

Mudde, Cas. 2004. "The populist zeitgeist," in *Government and Opposition*, Vol. 39, Issue 4, pp. 541–563.

Mudde, Cas. 2013. "Are populists friends or foes of constitutionalism." The Foundation for Law, Justice and Society.

Radek, Miroslav. 2019. "Election program of political party Kotleba – People's Party Our Slovakia," in *University Review*, Vol. 13, No. 1, pp. 46–50.

Sareen, Siddharth, Harriet Thomson, Sergio Tirado Herrero, João Pedro Gouveia, Ingmar Lippert, Aleksandra Lis. 2020. European energy poverty metrics: Scales, prospects and limits. *Global Transitions*, Vol. 2, pp. 26–36.

Index